Palgrave Studies in the Theory and History of Psychology

Series Editor
Jack Martin
Simon Fraser University
Burnaby, BC, Canada

Palgrave Studies in the Theory and History of Psychology publishes scholarly books that use historical and theoretical methods to critically examine the historical development and contemporary status of psychological concepts, methods, research, theories, and interventions. The books in the series are characterised by an emphasis on the concrete particulars of psychologists' scientific and professional practices, together with a critical examination of the assumptions that attend their use. These examinations are anchored in clear, accessible descriptions of what psychologists do and believe about their activities. All the books in the series share the general goal of advancing the scientific and professional practices of psychology and psychologists, even as they offer probing and detailed questioning and critical reconstructions of these practices.

More information about this series at
http://www.springer.com/series/14576

Heather Macdonald • David Goodman • Brian Becker
Editors

Dialogues at the Edge of American Psychological Discourse

Critical and Theoretical Perspectives

palgrave
macmillan

Editors
Heather Macdonald
Lesley University
Cambridge, MA, USA

David Goodman
Boston College
Chestnut Hill, MA, USA

Brian Becker
Lesley University
Cambridge, MA, USA

Palgrave Studies in the Theory and History of Psychology
ISBN 978-1-137-59095-4 ISBN 978-1-137-59096-1 (eBook)
DOI 10.1057/978-1-137-59096-1

Library of Congress Control Number: 2017947910

Cover design by Fatima Jamadar

Printed on acid-free paper

This Palgrave Macmillan imprint is published by Springer Nature
The registered company is Macmillan Publishers Ltd.
The registered company address is: The Campus, 4 Crinan Street, London, N1 9XW, United Kingdom

Acknowledgments

This volume would not have been possible without Jack Martin, the series editor for the Palgrave "*Studies in Theory and History of Psychology*," who has been a close colleague, mentor, and advisor. We are indebted to his unflinching clarity of thought regarding the discipline of theoretical psychology and to his generous wisdom. We also want to acknowledge the individuals who interviewed the selected scholars within this book. They gave their time and energy to this project with great enthusiasm. We are very grateful to Mark Freeman, Jack Martin, Elizabeth Corpt, Peter Maduro, and Marie Hoffman for their participation in the creation of these dialogues.

Many of these interviews took place at the Psychology and the Other Conference in 2015 at Lesley University. The beautiful Brattle campus has been the home of the *Psychology and Other Conference* since 2011 and we are indebted to this ongoing partnership and support. We also want to thank the wonderful group of students, friends, and colleagues that helped organize the conference and made it an ideal host for this critical interview series. Cacky Mellor, Sara Carabbio, Jacqueline Aug, Sam Gable, Danielle Moreno, David House, Ben Arcangeli, Abigail Collins, Adeline Dettor, Perah Kessman, Jenny Esdale, Kimm Topping, and Katie Howe-Goodman all exemplified selfless service. A special thanks to Kyle Bohl and David H. who helped keep the technology on track during these interviews.

A major debt of gratitude also goes to many members of the *Society for Theoretical and Philosophical Psychology* (APA, Division 24) for their participation as scholars in the conference and for being formative conversation partnerships.

Contents

Dialogues at the Edge

Heather Macdonald, David Goodman
and Brian Becker

> *But psychology is passing into a less simple phase. Within a few years what one may call a microscopic psychology has arisen in Germany, carried on by experimental methods, asking of course every moment for introspective data, but eliminating their uncertainty by operating on a large scale and taking statistical means.*
>
> —William James, The Principles of Psychology (1890/2007), vol. 1, p. 192

The original version of this chapter was revised in order to correct an error regarding the publication sequence and date of Psychoanalytic Treatment: An Intersubjective Approach (Robert Stolorow, 1987).

H. Macdonald (✉) · B. Becker
Lesley University, Cambridge, Massachusetts, USA
e-mail: hmacdona@lesley.edu; bbecker@lesley.edu

D. Goodman
Boston College, Chestnut Hill, Massachusetts, USA
e-mail: david.goodman@bc.edu

H. Macdonald et al. (eds.), *Dialogues at the Edge of American Psychological Discourse*, Palgrave Studies in the Theory and History of Psychology, DOI 10.1057/978-1-137-59096-1_1

Introduction

William James lived at the boundaries and did his best work there. In 1890, James published his magnum opus, *The Principles of Psychology*, a foundational work in which he makes clear distinctions about the kinds of assumptions each discipline takes for granted and then subjects these assumptions to critical examination. For many years, this was a widely used text that uniquely blended his personal reflections with his ideas on physiology, philosophy, and psychology. Throughout the book and his career, James extravagantly crossed the boundaries of multiple disciplines. This volume is comprised of a series of in-depth interviews with ten North American scholars and clinical practitioners who have similarly spent their lengthy careers working at the interdisciplinary edges of mainstream psychological discourse.

One of the criteria considered in selecting the thinkers in this volume, aside from their being situated in North America, is that each of them in their own scholarship has demanded a reinterpretation of thresholds that allows for a less monological emphasis in the adoption of a particular paradigm, and they have called their particular field to witness the historical, social, economic, and political consequences of their chosen frameworks. More specifically, these authors speak to the relationship between critique and creation and reflect on some of the most fecund sites for generating theory as well as on how these ideas can be applied to the discipline at large.

In order to structure these dialogues, the interviewers were sent over-arching, core questions to ask each scholar prior to the interview, along with a further set of questions (developed by the interviewers) that might speak to the particulars of the scholar's unique contributions to the field. Each interviewer has intimate knowledge of his or her interviewee's corpus of work, which allowed for deeply engaged conversations. The core set of questions included the following:

(a) How has the field of psychology responded to your work?
(b) What have been the resistances and impediments to your work being received in the contemporary discipline of psychology?

(c) If you had to speak prophetically, what would the field need to do to right its course?
(d) What are the historical, political, and/or theoretical concerns in psychology today that you feel have the most influence on the future of contemporary psychological practice?
(e) Historically, the practice of psychology has not associated itself with any specific forms of political activism or social responsibility. In your work, how have you responded to this issue?
(f) As you see it, what have been the crucial changes to your way of thinking or approach to the discipline over time?
(g) What do you imagine will be the future directions of your thought?

The majority of these conversations took place at the Psychology and the Other Conference held in Cambridge, Massachusetts, in the fall of 2015. The foundational intention for this conference was (and continues to be) to generate meaningful dialogue and exchange between psychologists, philosophers, anthropologists, sociologists, social workers, and theologians related to the ethical intersection of the self and the other. Attended by over 400 psychologists, philosophers, and scholars in the fields of theology and religious studies, this conference brought together perspectives that are frequently isolated from one another institutionally and historically. The conference deeply embraces the Socratic art of conversation and dialogue, which includes the idea that understanding is a process of co-creation between partners and that the subjective self is also realized through speaking (Bowie 1991).

Many broad themes were readily apparent in the responses to the above questions. Most of the scholars in this volume clearly suggested that one way for the discipline to "right its course" is to increase its capacity for interdisciplinary exchange. Drawing from a wide variety of theoretical traditions such as continental philosophy, relational psychoanalysis, and post-colonial theory would open up possibilities for creative solutions and new languages as a response to the challenge of human suffering. In the broadest sense, these are thinkers who have defined their work from theoretical traditions that speak to the indissoluble relationship between a person and their world and who argue for a selfhood that is relationally dynamic and historically situated.

Another theme that emerged is the idea that psychological discourse is an ideological category of knowledge and power, and the scholars suggest that some of the primary challenges within the field are associated with how rigid, objective scientific claims are produced, interpreted, and deemed to be universal truths. Questions regarding "truth and method" in the discipline of psychology are highlighted in many of these conversations as the problem of objectivity in experience, the problem of the relation between language and experience, and the problem of truth-making within dialogical communication. Certainly, the questions of truth and self-understanding in therapy are also measured against these kinds of analyses. Thus, what is central for these thinkers is the idea that language is always more than a statement; language is living and constitutes both the subject and the world (Gadamer 1960/1989).

However, in our view, what makes this group of authors most unique is that their work takes place at the edge of the conceptual and linguistic frames that currently set the theoretical boundaries of the discipline. The notion of the "edge" or "boundary" is important for understanding the context of these dialogues. A boundary is something that is in perpetual motion, not merely a line where the notions of "here" and "there" exist in binary opposition to one another. A boundary is a space of crossing that includes depth and height as well as horizontal shifts and expansions (Clingman 2009). Thus, these interviews highlight both the view of the field from the perspective of each thinker and also what each person has experienced as they have worked within or outside these spaces of crossing.

We believe that those who live at the frothy, turbulent edges of mainstream psychology and challenge the most cherished theoretical frameworks and methods of the discipline deserve engagement so that their works are brought into a wider view. Through the conversations with these scholars, we are able to access two parallel hermeneutics: their thoughts about their own scholarship and a wider perspective of the critical and creative fault lines and fissures that have shaped the discipline at large. Challenges to the psychological sciences that delimit what can be studied about human experience, a call for a fuller embrace of our encounter with difference, alternative methods and theories for inquiry, and social constructions of selfhood are just a few of the themes that permeate these conversations. In addition, these dialogues chronicle the responses of the scholars with regard to the current states of their respective areas of the field of psychology, their understanding of how their work has been metabolized, and the historical, political, or theoretical concerns in psychology today that they think will have the most influence on the future of contemporary psychological practice.

It could be said that the group of people gathered in this volume are within the bounds of postmodern and poststructuralist discourse. However, in each interview, the reader will see that this depends on where one draws such boundaries. There are myriad ways in which to define postmodern thinking, but it seems that there is at least a postmodern "mood" or tone around which these scholars appear to cluster. The shape and texture of such a mood is based on several axes upon which this kind of work at the edge seems to turn. The first is the idea that reality is not always pre-given and that the way the world shows up for one person does not reflect how it may appear for another. The second axis has to with the idea that meaning and language itself are context dependent, which challenges our notions of representation. Lastly, there is an ongoing suggestion or call for a methodological or perspectival pluralism that includes Aristotle's original definition of rationality, which involves the capacity to take multiple views at once without bias toward one or another. All of these axes relate to one another in a way that creates a type of seismic pressure, moving the flow of these scholars' discourse in directions that, at times, run counter to mainstream discourse. It is our hope that these dialogues will expose these points of friction as rich landscapes of thought and prospective places for future conversation.

The rest of the introduction provides a brief background to a number of theoretical and philosophical traditions from a variety of disciplines and explores how some of the scholars interviewed in this book take up or extend these traditions. This is not designed to rigidly place each scholar in a particular "camp" but rather to give examples of how crucial themes of discourse are highlighted in both the interviews and the scholarship of these thinkers.

Phenomenology, Hermeneutics, and the Interpretive Turn: Being, History and Dialogue

Man has no nature; what he has is history.
—José Ortega y Gasset (1973), *An Interpretation of Universal History*, p. 157

Many of the scholars in this volume draw on the philosophies of Martin Heidegger and Hans-Georg Gadamer as sources of inspiration for their work. Heidegger (1927/1967) and Gadamer (1960/1989) enlarged the scope of hermeneutics through the incorporation of phenomenology. The incorporation of phenomenology allowed the process of understanding and interpretation to be considered as an integral part of human experience. The significance of hermeneutics for the social sciences in general and for psychology in particular is a major theme that runs throughout these dialogues. Without giving in to extreme relativism or essentialism, many of the ideas expressed by the work of these thinkers assume that interpretation is central to both epistemology and ontology and to knowing and being. In fact, many are reluctant to separate the possibilities of existence from dialogical methods of understanding. Philip Cushman, Louis Sass, Robert Stolorow, and Brent Slife are authors whose scholarship has been in more direct conversation with the work of Gadamer and Heidegger (Cushman 2005; Sass 1992; Slife 2004; Slife and Christensen 2013; Stolorow 2011). A brief summary of the work of Heidegger and Gadamer may be helpful in order to summarize their distinctive contributions to the field as well as provide a solid contextual foundation for all of the interviews in this book.

The main purpose of Heidegger's (1927/1967) *Being and Time* is to ontologically explore the question, "What is the meaning of being?" Heidegger performs a phenomenological analysis of Dasein ("Being-there") that reveals the basic structures of human existence. Through a primordial examination of Dasein, he is able to establish the *existentiell* possibilities for a human being as such. For Heidegger, human reality is never complete; we are always more than we are, as existence pertains to possibility. This is a crucial concept embraced within each of these dialogues—that we as humans are never complete, and therefore there is no system that can grasp the totality of human behavior. Heidegger helps us get at the challenge of human existence a bit differently. He aspires to know, What does Dasein understand about its own being and Being in general? He asks, "What is the kind of being of the entity in which the world is constituted?" (Manning 1993, p. 23). Moving forward from these questions, Heidegger describes the open region, a clearing in which both the subject and object stand out. Existence, according to Heidegger, does not have exterior attributes but makes any kind of exteriority a dimension of Being. The world, then, is a rich, meaningful context in which we play out our most practical lives. For Heidegger, the world is the horizon in terms of which we understand Being, and the most prominent horizon in which we can comprehend ourselves is the horizon of time. Heidegger asks, What kind of time does Being have?

According to Heidegger (1927/1967), philosophers within the traditional discourse of philosophy asked misguided questions about Being, and they misconceived time by giving too much importance to the present. Thus, one of the goals in *Being and Time* is to provide a future-oriented analysis of time. Heidegger saw Dasein's involvement with the world as being-ahead-of-itself and defined the future as Being "toward oneself" (p. 377). In this way, Heidegger constructed an existential kind of time that does not overemphasize the present. Living in the moment is artificial because one can never be free of the past. We live our lives with respect to the future because the future is based on possibility and the past only has possibility based on the future. In other words, the future is lived by stepping out from the past.

In an attempt to bring out the temporal implications of existence, Gadamer draws on and expands Heidegger's *Being and Time*. In Part II of *Truth and Method*, entitled "Foundations of Hermeneutical Experience," Gadamer 1989 turns to the issue of "historicality" and the fact that understanding is always historically situated. Gadamer argues against the notion of pure objectivity and points to Heidegger's idea of *forestructure* to suggest that we cannot escape our prejudgments (prejudice), which are largely based upon the context of our past experience. According to Gadamer, bracketing our prejudices and views of our situation is impossible. However, what is possible is genuine dialogue. In genuine conversation, what happens in the process of understanding does not rest on a person's ability to enter the subjective field of another but rather is a matter of how the other *exposes* the subject matter at hand (*Sache*) that calls for understanding and dialogue in the first place. Understanding, then, is not a form of conceptual agreement between two minds; it is not any sort of conceptualization or determination of meanings. Rather, if the dialogue is successful, it will not be a prior reconstruction of prior meaning; it will *present* meaning. Understanding is the *production* of meaning.

The notion of tradition is at the heart of Gadamer's overall analysis. Understanding is to place oneself within a process of tradition in which we make judgments through a "historically effected" consciousness and a nexus of moral fabrics. Already embedded in webs of meaning and significance from tradition (and prejudice), we become moral agents of the traditions of the world from which there is no escape.

In his interview, Cushman acknowledges that his ideas regarding the cultural and social history of psychotherapy emerged in part from reading Heidegger's and Gadamer's analyses of historical embeddedness. He also discusses the importance of Michel Foucault's (1970/1994, 1973) notions of disciplinary power but is quick to point out that the exercise of power occurs within a tradition and within a network of moral understandings. So, for Cushman, Gadamer is able to rectify some of Foucault's cynicism. Gadamer's embrace of dialogue assumes that interpretation does not proceed on the basis of a sovereign attitude, such as the proposed metaphysics of psychic structures that guide some classical models of psychoanalytic interpretation.

Louis Sass also discusses phenomenological and hermeneutical themes as they have applied to his work with psychopathology, specifically schizophrenia. Sass has consistently argued for a phenomenological approach in the study of self, subjectivity, and the experience of schizophrenia. He asserts that these phenomenological approaches to psychopathology can be both explanatory and descriptive. He reminds us that Heidegger's analysis of the ontological dimension and his description of how the world may manifest for Dasein, goes to the heart of subjectivity; they are not disconnected or separated from one another.

In his interview, Robert Stolorow describes his realization that Heidegger's 1927/1967 existential philosophy, as presented in *Being and Time,* provides an invaluable philosophical tool for psychoanalytic phenomenology and the existential significance of emotional trauma. In particular, he cites

Heidegger's analysis of anxiety and being-towards-death as crucial insights in our understanding of psychological trauma. His work at the boundaries of psychology and philosophy led to his ideas on the psychotherapeutic encounter as a form of applied philosophy and phenomenology.

Brent Slife draws upon the hermeneutic approach to critique the assumptions of psychological science. In particular, he raises concerns regarding its individualism, naturalism, dualism, and bias against religious and spiritual traditions. Slife proposes, instead, a strong relationality that views the individual as inextricably embedded in a relational context. From his perspective, identity should be understood within its deeply textured historical traditions and communities. This gives rise to the importance of religion and spirituality as a common context that situates many individuals and cannot simply be abstracted as one variable among others. Although Slife is principally inspired by the hermeneuticist tradition (Heidegger, Gadamer, Ricoeur, etc.) he also understands his concept of strong relationality to be aligned with Levinas's ethical philosophy in which the individual's phenomenological encounter with the Other is always a radical moral calling.

Levinas and the Other: The Ethical Turn

> The Other faces me and puts me in question and obliges me.
> —Emmanuel Levinas (1961/1969), *Totality and Infinity*, p. 207

Another current or space of crossing that runs throughout the dialogues is a productive tension between the hermeneutic traditions, as they have been passed down by Heidegger and Gadamer, and a Levinasian phenomenology that addresses the inexorable ethical claim within interhuman experience. For some of these scholars, such as Mark Freeman (2014), the work of Emmanuel Levinas is an important inspiration.

Levinas is one of just a few philosophers who have challenged the imperialism of ontology and the primacy of knowledge garnered from an ontological analysis of Being. Levinas (1961/1969), in his book *Totality and Infinity*, conducts a phenomenological analysis of the ethical obligation to the Other and argues that "Truth is a modality of the relationship between the same and the other" (p. 290). He critiques Western philosophy as a whole, but he takes particular aim at Martin Heidegger and Edmund Husserl by contending that "ethics is first philosophy," not ontology. For Levinas, the idea of consciousness itself is based on the ethical relationship with the Other and not on the notion of an intentionality directed at an object or a Dasein that begins as a metaphysical presupposition for existence. For Heidegger (1927/1967), truth was an uncovering of entities from their hiddenness, or Being uncovered. For Levinas, any notion of so-called truth can only exist in the face-to-face encounter with another, since nothing can truly exist prior to relationships with others. The face of the Other is the primordial signifier and the origin of all discourse; the face summons one recognize its right to exist before any kind of metaphysics can be described. In this way, the project of knowledge-making and knowing is overturned by the ethical demand to respond to the demand of the Other.

Mark Freeman carries the ideas of Levinas even further by suggesting a whole host of ways in which the Other can offer unique forms of transcendence wherein our egocentric preoccupations are "arrested" and the "perimeters of the self" are not the limits of experiential and ethical possibilities. Music, nature, and the human face all have the capacity to pull one out of what he calls "ordinary oblivion," beyond oneself and more fully into the world and the experience it *gives*. Freeman describes this as a type of "thinking otherwise" where experience originates from outside of the prioritization of the self, not merely functioning as a "product of the psyche." Like Levinas, he wishes to prioritize the demand of the Other or even the Otherness of phenomena in our world that draw us forth.

Social Constructionism: A Related Affinity with Hermeneutical Thought

> Man lives in a world of meaning.
>
> —George Herbert Mead (1926), "The Nature of Aesthetic Experience," p. 382

Social constructionism, as established by such thinkers as George Herbert Mead (1934, 1938), Lev Vygotsky (1978), and Harry Stack

Sullivan (1953), shares an affinity with many of the hermeneutical traditions described above. Although not many of the scholars refer to it directly in their interviews, it certainly serves as an important undercurrent that flows throughout this collection. The theory stresses that throughout a person's lifespan there is an intersubjective arc of relational patterns, the meanings of which are embedded in sociocultural and sociopolitical matrices. The critical focus of social constructionism is to identify social roles and cultural positioning throughout the evolution of a person's life as a way to understand their individual psychological development and self-consciousness. The premise is that the coordination of social exchanges and the integration of different perspectives from these exchanges will reveal psychological insight into a person that does not exclude the cultural, political, or economic realities of their life (Gergen 1985). In fact, we are able to see directly how these are woven into their personhood through social or intersubjective acts. It is a perspective that follows the twists, turns, and oscillations of a person's life rather than the fixed geometry of chronological events.

It is important to note that Ken Gergen (1985) has been a champion of social constructionism throughout his career. Early in his scholarship, Gergen suggested that scientific knowledge generated by social psychologists influences the manifestation of whatever interactions or phenomena they were trying to describe. He asserted that science itself is nested in a series of social networks that determine the productions of its experiments and conclusions. In his later work, he focused more on the view of the "self" as purely relational and disabused the psychological profession of its notion of an individual "mind" through the argument that it is relational processes that shape the "mind" in the first place. The interview with Gergen emphasizes the case that he has made, in his more recent writings, for a "reflective pragmatism" which means shifting attention away from questions about "why some descriptions of human action [are] simply more accurate than others" to an examination of what values and meaning would serve society the best.

Through Gergen's scholarship, the idea of how to measure and describe human behavior goes to the heart of what these researchers reveal to us. He alerts us to the disproportionate gaps between experimental research (psychological science), theory, and practice, which other scholars also comment on in their interviews.

Challenges to Psychological Science

> I am not in my world. I am the boundaries of my world.
> —Wittgenstein, *Tractatus* (1922), p. xx

James Lamiell (2013) meticulously outlines the historical tensions between the disciplines of psychology and philosophy in a recent essay entitled "Psychology's Struggle for Existence: Appraising Wundt's Pessimistic Prediction a Century On." It is well known that Wilhelm Wundt is considered to be the father of experimental psychology, but few scholars have recognized that Wundt was deeply concerned about the separation of psychology from philosophy. Wundt was the first to open a laboratory and direct laboratory-based psychological dissertations with the assumption that consciousness could most appropriately be researched using the methods of the physical sciences. But Wundt also had conceptual concerns. Wundt, as quoted by Lamiell (2013), stated,

> Now, it is well known that when a married couple seeks a divorce, both members are usually at fault. In these pages it will be shown that the same is true in this instance, and that if this matter takes the course that both parties want, philosophy will lose more than it will gain, but psychology will be damaged the most. Hence, the argument over the questions of whether or not psychology is or is not a philosophical science is, for psychology, a struggle for its very existence. (p. 1)

These overarching historical tensions between experimental and theoretical-philosophical psychology are vividly mirrored in the interviews and dialogues that comprise this volume. Many scholars in the pages to come agree that what Wundt predicted has indeed come about—that psychology has been hurt the most—although not by what has been included within the mainstream discipline but by what has been excluded from it (e.g., qualitative methods, historical-interpretative methods, existential-phenomenological sensibilities, and social concerns). There has been an ongoing and tense "dialogue between empirical and theoretical understanding, between actuarial and clinical methods" (Mercer 2011, p. 1) in both research and practice. Lamiell in his interview also raises concerns around the distinctions "between statistical aggregates and the psychological lives of individual persons" (this volume) in order to highlight that aggregate data and the data of individual experience or narrative evoke different sets of scientific and philosophical assumptions.

In contemporary psychological discourse, the natural science approach is the privileged frame, upsetting the balance Wundt was attempting to maintain. This emphasis on scientific objectivity today can be seen in almost every standard textbook on psychology. However, influential philosophers, including Giambattista Vico, Johann Herder, Wilhelm Dilthey, and Friedrich Nietzsche, have offered alternative explanations of scientific psychological theories in order to suggest that science need not be the only arbitrator of our psychological questioning. Scholars like Hugo Munsterberg (1913) have argued that psychology is a qualitative pursuit and that psychology as a natural science and as a human science can, in fact, coexist and even complement one another.

In their interviews, James Lamiell, Ken Gergen, Mark Freeman, Louis Sass, Brent Slife, Robert Stolorow, and Phil Cushman and are all explicit in their call for a questioning of the deceptive certainty of the experimental sciences and for the discipline to recognize that science itself is culturally embedded. Gergen encourages a science that is valued less for its truth-telling and more for its practical capacity to contribute to society at large; Freeman suggests a need for a more "capacious and poetic" science and a science that maintains its fidelity to human experience; Sass invites a more sophisticated awareness of the correlates in subjectivity as they relate to neurobiological mechanisms; Slife questions the experimental methods frequently used to look at theistic claims and spiritual experience Stolorow suggests that we need to question the philosophical assumptions that dictate the type of evidence we look for in therapeutic outcome measures; and Cushman critically inquires into the proceduralism of experimentation as it is applied to therapeutic techniques and also shares his a significant concern about the loss of moral intersubjective frameworks for understanding one another.

Foucault and the Psychosocial

> Thought is no longer theoretical. As soon as it functions it offends or reconciles, attracts or repels, breaks dissociates, unties or re-unites; it cannot help but liberate or enslave. Even before prescribing, suggesting a future, saying what must be done, even before exhorting or sounding an alarm, thought, at the very level of its existence, in its very dawning, is in itself an action—a perilous act.
>
> —Foucault, *The Order of Things* (1970/1994), pp. 334–335

Another major challenge the scholars interviewed here offer to mainstream discourse is an understanding of the historical and cultural conditions under which disciplinary power, as defined by Michel Foucault (Foucault 1994), is produced as well as the effects of this power on the formation of psychological knowledge and clinical practice. In a lecture at Harvard University, Achille Mbembe (2016) summarized this idea:

> "Official culture" is the name for the process by which a ruling élite seeks to tame and domesticate its population by establishing official distinctions between the accepted and the unacceptable, the permitted and the forbidden, what can be seen and what cannot be seen, what can be represented and what cannot be represented. It is the process by which it coerces its subjects into internalising and reproducing truths not of their own making. (Public Lecture)

In their interviews, Derek Hook and Lynne Layton examine the ubiquitous presence of power and politics in the context of psychological processes that shape the very connections between the psychic and the social. Both thinkers discuss the ways in which the mechanisms of oppression have forged patterns of psychological and social practice that persist into contemporary larger social patterns and that bind or entangle the oppressor and the oppressed to memories of their past. In essence, they invoke political and psychological questions alongside one another by utilizing the revolutionary edge of psychoanalytic, post-colonial, and poststructuralist discourse (e.g., Frantz Fanon, Judith Butler).

Hook and Layton argue for a more sustained psychosocial engagement with the temporalities of overlapping and entangled histories that comprise psychological systems of knowledge. Through their scholarship, they observe the complex relationship between human desire and power as it continually forms and returns in the complex flow of political and psychological streams of human experience (Hook 2012, 2013; Layton 2006, 2009). The interviews with Hook and Layton offer the reader a critical engagement with crucial aspects of psychosocial discourse, including such issues as the racialization of space, remembrance, the body, human identity, oppression, structural violence, and mental illness, in order to explore the psychical mechanisms that undergird and substantiate these modes of reality.

Intersubjectivity and Relational Psychoanalysis

> "The subject" is no longer a substance endowed with qualities, or a fixed shape possessing dimensions, or a container awaiting the multifarious contents that experience provides: it is a series of events within language, a procession of turns, tropes and inflections.
>
> —Malcom Bowie, Lacan (1991), p. 76

As suggested in the previous sections, the authors interviewed in this volume embrace concepts of selfhood that are given meaning only in the context of others and in the world. They are inherently relational, interhuman, and intersubjective in their orientations. In particular, the scholarship of Robert Stolorow, Nancy McWilliams, Lynne Layton, and Phil Cushman has been instrumental in influencing the development of relational and intersubjective theories that have been very significant in the psychoanalytic world over the last several decades (Aron and Harris 2011; Cushman 1995, 2015; Layton 2009; McWilliams et al. 1999; Stolorow 1995). Their views of clinical or therapeutic discourse are infused with the understanding that human beings are *inseparable* from the interpersonal field. In their interviews, the readers will see the works of Harry Stack Sullivan (1953), Heinz Kohut (1971), and Donald Winnicott (1953) cited. These are clinical researchers and practitioners who placed an emphasis on personal meaning within the life of the individual and his or her relationships, where "meaning is to be found not in an objective, rational perspective, but in local, personal perspectives; the value of life is not measured by its conformity with a mature and transcendent vision, but by its vitality and the authenticity of its passion" (Mitchell and Black 1995, p. 169).

In addition, the early works of thinkers such as Margaret Mahler (1968), Donnel Stern (1985), and Jessica Benjamin (1988) provide additional dimensions to these conversations by incorporating attachment theory. They suggest that the ability to recognize the subjectivity of the Other is a key component to the internalization and development of object relations in infants and young children and that this idea also had enormous implications for the psychotherapeutic relationship. The emphasis on the differentiation and mutual recognition of selfhood that is so crucial in this phase of psychological development has become central to the development of "relational psychoanalysis" as a whole. The psychoanalytic encounter, from this perspective (and from that of an intersubjective theorist), is never "objective" or "neutral" but rests on the interaction of two subjectivities.

According to Benjamin's (1988) analysis, one cannot experience his or her own objectivity until one has recognized the subjectivity of the Other—the Other and the self become symbiotically dependent on one another in a process known as "mutual recognition." Mutual recognition blossoms into intersubjective attunement, which in turn blossoms into the excitement of being able to recognize, acknowledge, and celebrate shared feelings, allowing for a sense of vitality and authentic connection to life.

Robert Stolorow wrote his fourth book, entitled *Psychoanalytic Treatment: An Intersubjective Approach*, in 1987. He views the therapeutic encounter as fluid oscillations between the experiential worlds of the patient and the therapist. Multiple dimensions of experience oscillate between the background and foreground of each participant within the intersubjective field, and there is a mutual influence that creates the relational dynamic. In their more recent work, Stolorow and McWilliams include the pressures of global capitalism and larger societal forces as factors in the intersubjective encounter and argue that the subjective is layered in overlapping dimensions and contexts.

Nancy McWilliams, who is greatly influenced by the work of Stolorow, expresses the above point explicitly in her interview. She wonders if, as clinicians, we are able to view the patient through larger frameworks beyond the symptomology described in the *Diagnostic and Statistical Manual for Mental Health Disorders* (5th ed.; *DSM-5*; American Psychiatric Association 2013). She asks in her interview, "Is the person's life improving in the areas of love, in work, and play? Or increasing in self-esteem and affect tolerance and regulation or security of attachment?" And, "Do they have a sense of vitality?" Most importantly, she views these questions in the context of the intersubjective attunement she shares with each person she works with. As McWilliams says in her interview, "It's much harder in a mass culture, and now in a global culture, to feel like one matters at all, that one's efforts are going to be fruitful in some important way."

Lynne Layton and Phil Cushman also discuss the various structures of relational psychoanalysis and place them in conversation with larger social, political, and cultural processes to demonstrate that psychoanalysis often reproduces the very conditions that create the ills we wish to treat. Through their scholarship, Layton (2006) and Cushman (1995) expose the hidden moral and political discourses that live behind the technologies of the *DSM-5* and evidenced-based treatments and examine the impact of mass consumerism, neoliberal culture, and social media on the therapeutic relationship.

Conclusion

As we have suggested, it is our hope that these dialogues will expose disciplinary margins, boundaries, and crossings that will further contribute to already rich landscapes of thought and provide prospective places for future conversations. Because these representative scholars traverse the borders of multiple disciplines and simultaneously work at the margins of the mainstream discourse, they have a privileged perspective from which to comment on the discipline of psychology at large. Their acts of "transgression" are sites of contestation and a call to a more mature, capacious discipline in both research and practice. A mature discipline recognizes that selfhood is rooted in moral, political, and historical contexts and refuses anemic models of human experience that view potential and possibility in terms of production and consumption. A mature discipline also has the ability to reflect upon its discourse as a category of knowledge and power that shapes and produces subjectivity. In the political world, borders or boundaries can define zones of challenge and conflict, but they are also sites of social change and liberation. These ten conversations offer a specialized view from inside the fertile zone of creative critique.

Bibliography

American Psychiatric Association. (2013). *Diagnostic and statistical manual of mental health disorders* (5th ed.). Washington, DC: Author.

Aron, L., & Harris, A. (2011). *Relational psychoanalysis IV: Expansion of theory.* New York, NY: Psychology Press.

Benjamin, J. (1988). *The bonds of love: Psychoanalysis, feminism, and the problem of domination.* New York, NY: Pantheon Books.

Bowie, M. (1991). *Lacan.* Cambridge, MA: Harvard University Press.

Clingman, S. (2009). *The grammar of identity: Transnational fiction and the nature of the boundary.* Oxford, UK: Oxford University Press.

Cushman, P. (1995). *Constructing the self, constructing America: A cultural history of psychotherapy.* New York, NY: Addison-Wesley.

Cushman, P. (2005). Between arrogance and a dead-end: Psychoanalysis and the Heidegger/Foucault dilemma. *Contemporary Psychoanalysis, 41,* 399–417.

Cushman, P. (2015). Relational psychoanalysis as political resistance. *Contemporary Psychoanalysis, 51,* 423–459.

Foucault, M. (1973). *The birth of the clinic and archaeology of medical perception.* London, UK: Tavistock.

Foucault, M. (1994). *The order of things: An archaeology of the human sciences*. New York, NY: Vintage Books. (Original work published 1970).

Freeman, M. (2014). *The priority of the other: Thinking and living beyond the self*. New York, NY: Oxford University Press.

Gadamer, H. G. (1989). *Truth and method* (2nd rev. ed.). New York, NY: Crossroads. (Original work published 1960).

Gergen, K. (1985). *The social construction of the person*. New York, NY: Springer-Verlag.

Heidegger, M. (1967). *Being and time* (J. Stambaugh, Trans.). Albany: State University of New York Press. (Original work published 1927).

Hook, D. (2012). *A critical psychology of the postcolonial*. London & New York: Routledge.

Hook, D. (2013). (Post) apartheid conditions. London, UK: Palgrave.

James, W. (2007). *The principles of psychology* (Vol. 1). New York, NY: Cosimo.

Kohut, H. (1971). *The analysis of the self: A systematic approach to the psychoanalytic treatment of narcissistic personality disorders*. New York, NY: International Universities Press.

Lamiell, J. (2013). Psychology's struggle for existence: Appraising Wundt's pessimistic prediction a century on. *Journal of Theoretical and Philosophical Psychology*, *33*(4), 205–215.

Layton, L. (2006). Attacks on linking: The unconscious pull to dissociatiate individuals from their social context. In L. Layton, N. C. Hollander, & S. Gutwill (Eds.), *Psychoanalysis, class and politics. Encounters in the clinical setting* (pp. 107–117). New York, NY: Routledge.

Layton, L. (2009). Who's responsible: Our mutual implication in each other's suffering. *Psychoanalytic Dialogues*, 19, 105–120. doi:10.1080/10481880902779695

Levinas, E. (1961/1969). *Totality and infinity*. Pittsburgh, PA: Duquesne University Press. (Original work published 1961).

Mahler, M. (1968). *On human symbiosis and the vicissitudes of individuation. Vol. 1: Infantile and early contributions*. Madison, CT: International Universities Press.

Manning, R. (1993). *Interpreting otherwise than Heidegger*. Pittsburgh, PA: Duquesne University Press.

Mbembe, A. (2016, February 1). *The open laboratory: South Africa's Fanonian moment*. Cambridge, MA: Public lecture, Harvard University.

McWilliams, Mitchell, & Aron. (1999). *Relational psychoanalysis: The emergence of a tradition*. Hillsdale, NJ: Analytic Press.
Mead, G. H. (1926). The nature of aesthetic experience. *International Journal of Ethics*, 36(4), 382–393.
Mead, G. H. (1934). *Mind, self, and society from the standpoint of a social behaviorist*. Chicago, IL: University of Chicago Press.
Mead, G. H. (1938). *The philosophy of the act*. Chicago, IL: University of Chicago Press.
Mercer, B. (2011). Psychological assessment of children in community mental health clinic. *Journal of Personality Assessment*, *93*(1), 1–6. doi: 10.1080/00223891.
Mitchell, S., & Black, M. (1995). *Freud and beyond: A history of modern psychoanalytic thought*. New York, NY: Basic Books.
Munsterberg, H. (1913). *Psychology and industrial efficiency*. Boston, MA: Houghton Mifflin Co.
Ortega y Gasset, J. (1973). *An interpretation of universal history*. New York, NY: W. W. Norton & Co.
Sass, L. (1992). *Madness and modernism: Insanity in the light of modern art, literature, and thought*. New York, NY: Basic Books.
Slife, B. D. (2004). Taking practice seriously: Toward a relational ontology. *Journal of Theoretical and Philosophical Psychology*, *24*(2), 157–178.
Slife, B. D., & Christensen, T. (2013). Hermeneutic realism: Toward a truly meaningful psychology. *Review of General Psychology*, *17*(2), 230–236.
Stern, D. (1985). *The interpersonal world of the infant: A view from psychoanalysis and developmental psychology*. New York, NY: Basic Books.
Stolorow. (1995). An intersubjective view of self psychology. *Psychoanalytic Dialogues*, *5*, 393–399.
Stolorow, R. D. (2011). *World, affectivity, trauma: Heidegger and post-Cartesian psychoanalysis*. New York, NY: Routledge.
Sullivan, H. S. (1953). *The interpersonal theory of psychiatry*. New York, NY: Norton.
Vygotsky, L. S. (1978). *Mind in society: The development of higher psychological processes*. Cambridge, MA: Harvard University Press.
Winnicott, D. (1953). Transitional objects and transitional phenomena. *International* Journal of Psychoanalysis, *34*, 89–97.
Wittgenstein, L. (1922). *Tractatus*. New York: Harcourt, Brace & Company Inc.

Heather Macdonald is Assistant Professor of Psychology at Lesley University. She came to academia after years of practice as a clinical psychologist whose work involved community outreach, child assessment, and individual therapeutic services to children and families in the foster care system and with youth involved in the juvenile justice system. Her work in urban environments and abroad has led to scholarly research on the interface between culture, social justice, relational ethics, clinical practice, and postcolonial thought. Her first monograph, titled *Cultural and Critical Explorations in Community Psychology* (2017) published by Palgrave Macmillan, further considers the implications of psychological assessment, diagnosis, and historical trauma. As much as the book represents a cultural critique of more traditional clinical discourse, it also suggests that if we view clinical relationships from the perspective of a relational ethics, that includes discourse about cultural memory, freedom, power, and other events that break with the neoliberal expectation of "mutual reciprocity," then therapeutic work can also be a site of transformation and reversal that moves beyond fixed racial categories. She is a fellow of the Psychology and the Other Institute.

David Goodman is the Associate Dean of Academic Affairs and Advising at the Woods College of Advancing Studies at Boston College, Associate Professor of the Practice in the Department of Philosophy in Boston College's Morrissey College of Arts and Sciences, the Director of Psychology and the Other, a faculty member at the Massachusetts Institute for Psychoanalysis, and Teaching Associate at Harvard Medical School/Cambridge Hospital. Dr. Goodman has written over a dozen articles on continental philosophy, Jewish thought, social justice, and psychotherapy. Dr. Goodman currently serves as the Series Editor for the Psychology and the Other Book Series with Routledge. He has authored and edited several books including *The Demanded Self: Levinasian Ethics and Identity in Psychology* (with Duquesne University Press, 2012), *Psychology and the Other: A Dialogue at the Crossroad of an Emerging Field* (with Mark Freeman and Oxford University Press, 2015), *The Ethical Turn: Otherness and Subjectivity in Contemporary Psychoanalysis* (with Eric Severson and Routledge, 2016), *In the Wake of Trauma: Psychology and Philosophy for the Suffering Other* (with Eric Severson and Brian Becker and Duquesne University Press, 2016), *The Road to the Living God: Ana María Rizzuto and the Psychoanalysis of Religion* (with Martha Reineke and Rowman & Littlefield, in press), and *Memories and Monsters* (with Eric Severson and Routledge, in press). Dr. Goodman is also a licensed clinical psychologist and has a private practice in Cambridge, Massachusetts.

Brian Becker is Associate Professor of Neuropsychology and Assistant Chair of Psychology at Lesley University. He is a Research Fellow at the Psychology and the Other Institute and co-chair of its biannual conference. His research explores the implications of recent French phenomenology for psychological theory and practice. He is co-editor of *In the Wake of Trauma: Psychology and Philosophy for the Suffering Other* (Duquesne University Press, 2016), and has most recently published in the "Journal of Theoretical & Philosophical Psychology and Humanistic Psychologist." Dr. Becker is a licensed clinical neuropsychologist.

Strong Relationality and Hermeneutic Realism: A Conversation with Brent D. Slife

(Interviewed by Brian Becker, Lesley University)

Dr. Brent Slife is a practicing clinical psychologist and the Richard L. Evans Chair of Religious Understanding and Professor of Psychology at Brigham Young University. He has been honored with several awards for scholarship and teaching including the Presidential Citation from the American Psychological Association for his contribution to psychology; the Eliza R. Snow Award for research on the interface of science and religion; the BYU Alumni Professorship Award for Outstanding Teaching and Scholarship, and the Psi Chi Most Outstanding Professor Award. He served as the President of the Society of Theoretical and Philosophical Psychology, on the Council of the American Psychological Association, on the editorial boards for multiple journals, as a Routledge book series editor for *Advances in Theoretical and Philosophical Psychology*, and currently is editor-in-chief for the *Journal of Theoretical and Philosophical Psychology*.

B.D. Slife (✉)
Brigham Young University, Provo, UT, USA
e-mail: brent_slife@byu.edu

H. Macdonald et al. (eds.), *Dialogues at the Edge of American Psychological Discourse*, Palgrave Studies in the Theory and History of Psychology, DOI 10.1057/978-1-137-59096-1_2

He has authored or coauthored over 220 articles and ten books. The titles of some of those books include: *Taking Sides: Clashing Views on Psychological Issues* (2013) now in its 18th edition and used across the entire country to help students learn how to critically engage the field of psychology; *Critical Thinking about Psychology: Hidden Assumptions and Plausible Alternatives* (2005), which was an APA best seller; *Critical Issues in Psychotherapy* (2001); and *Time and Psychological Explanation*. His newest books, both published in 2017, are entitled *Frailty, Suffering, and Vice: Flourishing in the Face of Human Limitations*, published by APA Books, and *The Hidden Worldviews of Psychology's Theory, Research, and Practice,* published by Routledge.

Throughout these works, Dr. Slife has rigorously and relentlessly examined the theoretical underpinnings of psychotherapy and scientific methodology including psychology's implicit assumptions of linear time, dualism, naturalism, and liberal individualism, just to name a few of the many philosophical issues he has tackled within psychology.

Beyond these critical analyses, Dr. Slife has explored constructive theoretical alternatives for psychology, drawing upon hermeneutics, proposing a relational ontology that attempts to shift away from psychology's penchant for individualistic approaches to human subjectivity. His research also looks into ethics, virtue ethics, and definitions of human flourishing.

Finally, his work has also addressed difficult and not so easily reconcilable relations between theology, spirituality, and psychology, asking, for example, whether belief in God matters to how we theorize, research and practice psychology.

Brian W. Becker (BWB): I want to start off on an historical note, to get a sense of how you got involved in theoretical psychology. Your mentor, Joseph Rychlak, was a student of George Kelly actually, so that makes you George Kelly's grandson, I guess?

Brent D. Slife (BDS): Yes, it does. Very proud of that.

BWB: It's a wonderful lineage. You have previously proclaimed your mentor to bc the Kant of modern psychology, and one of Rychlak's major contributions was his critique of modern psychology's over emphasis on material and efficient causality. Rychlak, instead, sought to refocus the emphasis back on final and formal causality. What influence did your mentor and his ideas have on your involvement and philosophical orientation toward theoretical psychology?

BDS: Marvelous question. If I could pause just to thank you, Brian. I've mentioned in our email exchange just how thoughtful you have been about reading my material and how generous your introduction was. Given that our mechanical breakdown delayed our flight, and so forth, I'm just thankful to the audience, and particularly Heather and David and our cameraman, of course, for getting up earlier and you guys sticking me in the conference here somewhere. I'm very grateful.

So yes, I am proud to be an intellectual grandson of George Kelly. He and my mentor, Joe Rychlak, are some of the clearest examples of Kantians in psychology. Their attention to the pivotal role of the person's interpretation in experience was crucial to my eventual move to a hermeneutic framework. Less known, perhaps, was their advocacy of the dialectic. Joe's not particularly Hegelian, but he and Kelly did recognize the import of contrasting meanings, perhaps even "otherness," in the very definition and identity of meanings and experiences. For example, one cannot meaningfully judge people as "beautiful" without knowing what they might be like if they were "ugly." Even the profile or boundary of an object requires knowledge of both what it is and isn't to fully comprehend it. Kelly was one of the first to note this pivotal aspect of all meanings. They aren't just related to their similarities; they're no less related to their contrasts.

Kelly also realized the significance of a dialectical reasoning ability in what Rychlak would call free will, I would call human agency. A dialectical reasoning ability is necessary as a kind of option generator; it makes possibility possible. In judging someone as "good," for example, the dialectic also allows the judgment of "bad" to be possible, with a range of possibilit ies in between, because the meanings of good and bad aren't completely separate meanings in this sense. To decide to turn to the right is also to have decided not to turn to the left, allowing us to have true options—the *otherwise* of a human agency.

Brian, you're also right that Rychlak influenced me in combining this type of Kantian outlook with a more Aristotelian framework. Most notably, Rychlak (1981) saw human nature in an Aristotelian final causal or teleological framework, with us all being essentially goal- or meaning-oriented creatures (Slife and Williams, 1995). Joe also used Aristotle's formal causality, more of a holistic understanding, to understand this goal-oriented nature within a culture. Because the person is seen as inherently goal- or even future-oriented, I felt this view necessarily begged ethical questions, such as: which goal or future is best? Which goods should I, as a human being, strive for? And the answer to these questions is the reason I have been interested in virtue ethics and even theistic worldviews.

Even before my Rychlak influences, I was interested in more theological and philosophical questions and became an atheist through high school and the first part of college. Philosophy was something I could use to out-argue my Southern Baptist friends. And perhaps ironically, philosophy brought me back to religion, especially when I realized that it was possible to think deeply, even intellectually, about religion. In this sense, Joe Rychlak was a model for me because he was a staunch Catholic who was also a critical thinker of the first order.

BWB: One of the major theoretical considerations addressed throughout your writing and I want to explore with you today is the question of relationality. It's almost become a fad now in many parts of the field, where now different psychological models, including neurobiology, emphasize relationality as constitutive of the person. And, of course, at the other end of the spectrum there's Gergen's (See this volume) social constructionist versions of relationality. You've examined in your own work different kinds of relationality, namely a "weak relationality," which you've identified in some versions of psychology and what you're proposing as a "strong relationality." Could you discuss the distinction between those terms and how it overall ties in with your overall thinking about what you call a relational ontology?

BDS: Yes. Strong or ontological relationality has a fairly basic claim: that the context and the person are mutually constitutive of one another (e.g., Slife, 2004a). They have, in an important sense, a shared being, and I certainly think that this kind of thinking is reflective of much work at this conference. The foil of ontological relationality—and I think all theoretical frameworks have a dialectical other, or foil—is abstractionism. An abstractionist ontology, by contrast with strong relationality, includes the basic assumption that things and events are most real when they are divorced or abstracted from their contexts. Contexts, such as surrounding environments or history, are viewed as confounding variables that should be factored out to understand the thing or event. The laboratory tradition in science is a good example of this abstractionist tradition. Things are better understood when they're sterilized from the other things that surround or went before (or after) them, which relationality would completely dispute.

The abstractionist tradition also champions abstractions as our highest form of thinking. Abstractions have any number of manifestations, from theoretical propositions to universal principles to religious beliefs to mathematical formulas. They are often viewed in Western culture and Western science as the pinnacles of knowledge, whereas the strong relationist sees all these abstractions as helpful, but ultimately too thin and reduced to be the ultimate goal of knowledge. To really be useful, according to strong relationality, is for knowledge to be situated and understood as ultimately related to practices.

With that modicum of background into the two main ontologies—relational and abstractionist—I can now attempt to answer your question about the difference between strong and weak relationality (Slife and Wiggins, 2009). Put perhaps too simply, the ontological relationist sees most relationships as constitutive relationships, parts of wholes, where the parts not only cause the whole, but the qualities of the parts themselves stem from their relation to one another (the whole). The very identity or properties of a part, in this sense, come from its relationship to the other parts. The parts have, in a real sense, a shared being.

Weak relationists, on the other hand, do not typically understand connections as constitutive. They may care deeply about relationships of all types (e.g., interpersonal, interactive, systems), but they view the relationships among the parts of the system differently. First, they interpret the parts as fundamentally separable from other parts. We see this a great deal when theorists explain people from a liberal individualist perspective. Persons are first and foremost individuals from this perspective, with all of their necessary properties and qualities contained within them, including their biological properties and their psychological qualities. Secondly, these self-contained individuals can influence one another, but they do so across the time and space that separates them, leading ultimately to a dualism of the subjective person and the objective surrounding environment. For a person to be influenced from a weakly relational (and liberal individualist) perspective, we would have to get the "outer" information "inside" the person, inside the mind which is itself seen as internal.

With strong relationality, by contrast, there is less emphasis on the inner and the outer. Minds are not necessarily "inside" of us, for example, because they are always and already situated within our socio-historical context—with parts of our mental memories, for instance, in our phones and our computers. Our interpersonal relationships also loom quite large, because they are literally part of us, part of our personalities. I am a node or nexus of relationships—a father, cousin, husband, interviewee, etc. The quality of our interpersonal relationships is vitally involved in the quality of our individual character, which recent developmental research is now discovering. A maturation process is not a movement from dependence as an infant to independence as an adult, with dependence vilified in such abstractionist or individualist concepts as Dependent Personality Disorder or co-dependence. Maturation and true development are, instead, a movement from one type of dependence to another. We are always dependent, and self-sufficiency is illusory (Fowers, Richardson, and Slife, 2017).

Strong and weak relationists also view the connections or unity among community members differently. Weak relationists assume that abstractions, such as the beliefs of a religious community or the philosophy of a political party, bring people together and ultimately unify them. This viewpoint—unfortunate from a strong relational perspective—puts too much emphasis on similarities, with many of these communities presuming that the commonality of abstractions (e.g., beliefs, principles, propositions) are what attract and build unity, with differences and otherness as an ultimate threat to such groups (Slife, 2004a). Dating websites such as eharmony.com and match.com exemplify this emphasis on similarities in a marital "community."

Strong relationists, however, consider relational differences to be as important to the unity of communities as relational similarities, just as the dialectic with meanings and parts with wholes. For example, I can say without hesitation that one of the many strengths of my marriage is our differences, which are at least the spice of our marriage, and perhaps more substantively what I delight most in—Karen's many intriguing (to me anyway) differences from me. Because differences are less feared in strongly relational communities, conflicts and tensions are expected and seen as ultimately enriching the community rather than something to be feared or viewed as ultimately threatening. This understanding in strongly relational communities doesn't mean that just any difference (or similarity) will strengthen relationships or unify people. The strong relationist knows that we are also constituted by the contexts of our moral traditions (unlike the liberal individualist) and can thus use these traditions as resources for understanding what differences are good and right (e.g., virtue ethics, religious traditions).

In any case, I hope that gives you a bit of an idea of how weak and strong relationships are different (and similar).

BWB: What you said reminds me, on the political level, of Michael Walzer's concept of thick and thin, where there is this modernist belief that groups of differing cultures can be sufficiently motivated and unified by a universal, common ethics that binds all people together. However, Walzer (1994) argues, it is rather the contextualist, thick understanding of one's moral discourse that actually moves one to action. And so I see a similar parallel between this and what you're discussing in regards to psychological science. Is that a fair comparison?

BDS: It is a fair comparison. Walzer's (1994) concepts of thick and thin clearly correspond to aspects of the two ontologies I'm describing here. The universal principles you are referring to are ultimately abstracted from their contexts, or the contexts would change them and they couldn't be universal or principled. This abstractionism also implies that the modernist in Walzer's account will ultimately extol some kind of similarity to the principles or beliefs as a test of cultural unity, and identify differences as cultural threats. Such threats are frequently considered outliers or even infidels and thus need to be eliminated in some manner to save the group's unity (e.g., ostracism, terrorism).

I don't remember Walzer that well but in strong relationality the moral discourse that unites a community would be viewed as necessarily varied and textured. Similarities are never unimportant, but the strong relationist would expect rich cultural differences, given the varied moral traditions that could interpenetrate virtually any culture. If you asked, for example, a member of a Western religious group what unites them, they would likely provide an abstractionist reply. A set of propositional beliefs is often cited as if the similarity of belief is the main source of member unity.

A strong relationist, on the other hand, would predict that this unity of abstractions is ultimately illusory, because any reasonable inquiry into a member's individual beliefs, even within a devoutly religious community, would reveal significant uniqueness of belief and thus many differences with other members (though they may profess a belief in the common abstractions). The moral discourse that actually unites communities, then, has to be thick rather than thin and complexly textured rather than abstractly reduced to a set of principles. From a strongly relational perspective, the moral discourse that really brings the culture unity (e g., a textured narrative) is the discourse of good and virtuous relationships within and outside the community (e.g., with their God). Good and virtuous relationships can be determined by any relational morality, such as virtue ethics or some forms of theistic ethics.

BWB: Could you help clarify the connection between your use of the word strong relationality and your use of the term relational ontology? How are those two terms connected?

BDS: I use strong relationality and ontological relationality interchangeably, with "strong" used in the philosophical sense of "hard" (e.g., hard determinist) or perhaps even more extreme. I certainly do not mean to imply that weak relationality is bad or wrong, but rather it's a softer or less stringent brand of relationality.

BWB: And can weak relationality also represent another relational ontology?

BDS: No. What I'm calling weak relationality stems from an abstractionist ontology. Weak relationships are the only relationships that an abstractionist or individualist can conceive of, given their ontological assumptions of the ultimate separability of parts and wholes, persons and their communities, individuals and their traditions, present and past. In other words, I don't doubt for a minute that many abstractionists, even if they don't realize they're abstractionists, genuinely value and care deeply for interpersonal relationships. Therapists, for example, have long held that it's the relationship that heals, but the question here is what does relationship and healing mean (Slife and Wiggins, 2009). Is it the incorporation "inside" of "outside" abstractions (e.g., principles, beliefs) that heals, as in the weakly relational approach, or is it the shared being of the person and context in a manner that constitutes good character and cohesive community?

As a social psychologist friend of mine, Jeff Reber, would say, social psychologists aren't really "social" at all in the strongly relational sense, with a few exceptions. Strong relationality isn't apparently even available for them to use theoretically, which is a shame from my viewpoint. Instead, persons are automatically understood first as self-contained individuals and only later, after they interact, as social beings. Even after they've interacted, external influences must be brought "inside" the organism for them to be influential, whereas the strong relationist would assume that the outside cannot *not* be at least partly constitutive of the self. The context, including our body, one's history, a person's vision of the future, and the immediate situation, among other factors, is always and already part of who we are. I often tell my students that I couldn't be an instructor if they weren't in the class as my students. In other words, if I did what I usually do in class, including answering student questions, and no one was in the classroom with me, I'd be better identified as a mental patient rather than a patient instructor, and this identity would occur regardless of what I might incorporate "inside" my mind.

BWB: So let's switch to the topic of dualism since we've already started there. You've already suggested that dualism is a problem in some forms a weak relationality. What role does the issue of dualism play in your own work, and what is your critique of it, particularly as it applies to science.

BDS: Sure. I see the dualism of abstractionism, especially the dualism of subjectivity and objectivity, as pervading the work of many psychologists (Slife, Reber, and Faulconer, 2012a). The widespread notion in research, for example, that bias is bad is, I believe, a by-product of the dualist notion that investigators are supposed to clear away all subjectivity, such as biases, to get to the pristine objective world. The dualism here is the simple assumption that the subjective can be separated from the objective. The strong relationist, by contrast, does not make this assumption. In fact, the relationist assumes one of several forms of nondualism, where the subjective and the objective cannot be separated, where our interpretations, values, and even biases are part and parcel of not only our experience of the world but also any form of knowledge.

What I believe that you and I, Brian, have discussed to a limited degree concerns the application of these dualistic assumptions to other types of relationality, such as social constructionism. As you know well, Ken Gergen is a leading social constructionist whom I have criticized for not being relational enough. His book, *Relational Being*, is a marvelous relational work at so many levels. Still, he is fundamentally only weakly relational because his brand of social constructionism contains dualist assumptions that ultimately separate the subjective, or the inter-subjective in his case, from the objective (Slife and Richardson, 2011).

My criticism might make more sense with a simple example. Consider the hypothetical of a careless male driver killing a little girl with his truck. Most witnesses of this incident will experience at least one meaning, the tragedy of the little girl's death. The question for us here is: what is the source of these meanings? The most likely answer to this question among psychologists would be two forms of dualism. The first is an individualist form of dualism, where the psychologist contends that the tragic meaning stems from the individual witnesses endowing the objective event with subjective interpretation. In other words, the source of the tragic meaning is the individual's subjectivity; the objective crash is not meaningful in this sense. An objective rendering of the crash would consider the force of the truck in relation to the physiology of the girl, etc., but this solely objective description is not supposed to consider the tragedy of a little girl's life cut short.

Now most social constructionists would be rightfully critical of this individualist account. Because the tragic meaning of the girl's death is socially constructed, that is, constructed culturally through our interpersonal relationships and interchanges with one another, the subjective account of the individualist has to be wrong. Indeed, Dr. Gergen claims expressly that this kind of social constructionism precludes the dualism of the individualist, because the source of the tragic meaning lies in the relationship of the people to one another—a kind of relationality. The witnesses in this case must have learned from their culture how to interpret the girl's mangled body. The source of their meaning, in this sense, is inter-subjective rather than subjective (Slife and Richardson, 2011).

Nevertheless, what is rarely discussed among social constructionists is that the objective outcome of the crash—the little girl's dead body—still contains no tragic meaning. This meaning stems solely from the negotiated (socially constructed) meanings of the culture, which are then imposed onto the body. The body itself is meaningless in this regard. I say it this way because the separation of the objective and inter-subjective, the essential dualism, is preserved. The only change is from the subjectivity of the individual to the subjectivity of the culture, each still providing the meaning for the mangled body. Even the morality of this meaning—that the driver shouldn't have caused this tragedy—is itself constructed in this manner. That is to say, the girl's dead body has nothing, in principle, to do with either the tragedy or the injustice of the crash. The girl's body is neutral or free of values altogether.

This is not a strongly relational account in my view, because it is not an account in relation to the complete context of the event. It is insufficiently relational because it takes account of the interpersonal context, the constructors of the culture, but it doesn't take account of the *im*personal context, the truck having killed the girl. It is true that the truck and the dead body are not constructors, in which case they cannot participate in the social construction of the tragedy, but it is not true that the *im*personal situation is not itself a contributor to the meaning. In other words, the girl's mangled body is itself a real meaning, a violation of a "form of life," to use Wittgenstein's term, or a "human constant," to use Taylor's phrase. The tragedy of a life cut short is not merely "created" or "invented" by the individual or the culture, to use some of the terms from Gergen's book; the mangled body makes its own contribution to the tragedy; it's at least partially constitutive of this tragic meaning (Slife and Yanchar, in press).

The strong relationist can readily admit that there are other contributors to the meaning, including the individual's own interpretation and the culture's shared interpretation, but the situation itself is also a pivotal contributor. There is a moral reality or even a truth to the situation itself, implying that the person who stands over the mangled body and judges it to be "delightful" is at least morally wrong, no matter how different their culture or unique their subjectivity. In the later Wittgenstein, to delight in a girl's dead body is to commit a mistake of grammar, because his notion of grammar is as normative as it is semantic or syntactic. As I've said to Gergen himself, if he and I were to hike up one of our Utah mountains, the steepness of the mountain can be interpreted in many different ways. However, this steepness cannot be socially constructed away; the impersonal mountain itself contributes to the meaning of our hike. For this reason, then, the social constructionist account is incompletely relational and is ultimately dualist in the classic Cartesian sense.

BWB: A lot of your work has dealt with hermeneutics and, part of what I heard you saying in your last response, to borrow from hermenuetical terminology, is that there's a horizon in which the person operates as they encounter such an event. How has hermeneutics influenced your own thinking in light of a non-dualistic strong relationality?

BDS: It's heavily influenced my thinking. I see everything I've been discussing with you as complementary to, if not completely consistent with hermeneutics, indeed, primarily inspired by hermeneutists such as Taylor, Gadamer, Heidegger, Ricoeur, and Richardson. Strong or ontological relationality itself is a hermeneutic ontology. Even the hypothetical of the mangled girl is probably best considered an explication of hermeneutic realism, where the meanings of the world are real. The term "real" here does not imply subjectivity-free, objective entities. Meanings are contextually and interpretively bathed, but they're real in and through that bathing, much as the witnesses to the tragedy of the girl would feel that the meanings are inherent in the crash, even though they don't occur without their cultural and individual interpretation.

This new, meaning-oriented understanding implies that the world consists of meanings rather than objects (Slife and Christensen, 2013). Psychologists, in the spirit of abstractionist dualism, have viewed their subject matter, such as humans, as natural science objects. Objects in this sense are conceived as though context doesn't matter, and so the objects themselves do not essentially change with changing contexts. Persons, for example, are considered to have personality traits that rarely change. This table is always a table, no matter the context. Also, objects are defined in terms of what they are, not what they aren't, so there is no difference or otherness involved in the qualities or identities of objects. The person, for this reason, is automatically assumed to be an essentially unchanging entity in which otherness is viewed more as antagonist rather than protagonist in the narrative of one's life.

BWB: I want to pick up on that theme of otherness for a moment given the nature of this conference as Psychology and *The Other*. Levinas, in particular, has strongly influenced many of the scholars at this conference. But there seems to be a tension I find, between, say, a Levinasian conception of "otherness" and the hermeneutic tradition as handed down from Heidegger (1927/1967) and Gadamer (1960/1989), where hermeneutics, according to Levinas (1960/1989), is not sufficient and, more importantly, it's not most fundamental. Rather, otherness is a primordial experience out of which hermeneutics later emerges. Do you perceive this tension between these different philosophies and how do you reconcile these approaches in your own work given both influences on your thinking?

BDS: You're right that I do find both the Levinasian and hermeneutic traditions inspiring my own work, so I certainly don't see them as enemies of each other. I might even make a case for them being somewhat complementary to one another, as I know my Levinasian friend, Ed Gantt, would agree.

Let's begin with the hermeneutic tradition where, as we've said, meanings are not only real and shared but also inherently value-laden. Given this nondualist tradition, as I've described, these inherent values are not imposed subjectively onto an objective reality, as if we are the controllers of the values. These values are, instead, part of reality itself, implying that like the rest of reality they can lay claim to us, perhaps even control us to some degree through our experience of them. They thrust themselves upon us (Slife and Yanchar, in press). We could even say in a Levinasian manner that they thrust themselves upon us from a height, with some mattering more than others. And surely no hermeneutist would argue with the probability that this moral realism, this very real ethic, is likely most poignant and significant in regard to our interpersonal relationships, which Levinas, of course, fleshes out through the particularity of the person's countenance, etc. As the hermeneutist Svend Brinkman (2011) contends, it's the claim of morality or ethics on each of us that forges the starting point of the practical hermeneutic—a contention Levinas might find compatible.

As for the otherness of the Levinasian Other, of the person's face, for example, I think I've already explained how meanings from a hermeneutic perspective have an inherent dialectical quality. Meanings are never just what they are, and they aren't related to other meanings solely through their similarities. No, they also reach out to and require for their very identities what they're not, the otherness of themselves. As my old mentor, Rychlak (1981), would put it, meanings "reach beyond themselves," they're overdetermined, so otherness is expected and badly needed. Again, it also makes sense to me, an ontological relationist, that Levinas would single out interpersonal meanings in this regard, and so I'm not sure I see why there have to be tensions between the two traditions.

BWB: But to the degree that Levinas rejects ontology, Levinas' thought appears to be problematic for you if you want to maintain ontology as a fundamental concept in your own thought?

BDS: Well, part of a truly relational ontology is the relation among the meanings or categories of philosophy, such as the categories of ontology, epistemology, and ethics. I think far too often philosophers get caught up in which of these categories gets priority. In fact, to decide which category comes first is to make the assumption that ontology and ethics, as pertinent examples, are separable as meanings, and I don't really accept that these categories are separable. I would argue, instead, that all ontologies have ethical implications and assumptions, and all ethical frameworks have ontological implications and assumptions. Although the distinctions among such categories are practically helpful, it doesn't necessarily follow that these categories are completely separate from one another.

In the case of ontological relationality, there is no question in my mind that this ontology implies the moral value of virtuous relationships. I've written about this in several articles, and the ethic that stems from this ontology is one of the foundations of my own approach to psychotherapy. In fact, L Jay Mitchell and I founded a therapeutic boarding school for adolescent women that is based on this ethic (Slife, 2012). Similarly, I view Levinas as fitting nicely into the tradition of hermeneutic realism, where meanings are most fundamentally real and meanings are contextual, value-laden, etc. I may read Levinas differently from some others, because I would agree that he appears concerned with the priority of these philosophical categories in early works, such as *Is Ontology Fundamental* (1989)? However, I'm less sure that he views them as separable in later works, such as *Otherwise Than Being* (1998). Is it just possible that his main concern was the false prioritizing of ontology in philosophy, which I would heartily agree with, rather than the prioritizing of a totally separate ethic? Maybe my problem is that I read hermeneutics through Levinas (or the reverse), but whichever way it is, I view them as more complementary than contradictory.

BWB: Yes. So, you are rejecting some of the Platonic and later Cartesian ontologies that have been there?

BDS: Great point. A relational ontology, in the Platonic or Cartesian sense, is more of an anti-ontology. I tell my students its more anti-theoretical than theoretical, because the theoretical tradition in psychology is connected more to postulated universals that cannot allow for the changeability of context, abstractionism again. Sure, we're sitting here talking in quasi-theoretical terms about ontological relationality, but it's not the talk or the terms that comprise this ontology; it's concrete relations themselves. Strong relationists readily acknowledge the linguistic reduction of any description or articulation. Language is necessarily abstract and thin, to use Walzer's term, especially in comparison to the richness of practice or lived experience. However, relational ontologists do not take their *description* of concrete relational experience to be the fundamentally real. They take the rich and detailed practices and experiences *themselves* as the fundamentally real (Slife, 2004a).

BWB: So there are other, more fruitful, ontologies that don't abstract the way those philosophers did?

BDS: That's correct.

BWB: Speaking of the Other, theistic approaches in psychotherapy has been another theme in your work. You've engaged in some major critiques against psychology for the way in which its methods are biased against the world view of religious individuals, and you also propose some constructive means for how conversations between religion, theology, spirituality and psychotherapy may take place.

So, what are some of the ways in which you have thought about a theistic approach within psychology?

BDS: Yes, the philosophy of theism has attracted me primarily because it highlights, again, through a kind of dialectical otherness, the embedded naturalism of psychology—what Jamie Smith (2014) and Charles Taylor (1989) call its "immanent frame." Yes, I think that a naturalistic framework has captured American psychology and even Western culture. In fact, I see it across virtually all the disciplines of the Western academy. Naturalism is the taken-for-granted worldview that only natural events matter and only naturalistic explanations are valid (Slife, 2004b). This view is not something empirically derived; it's an assumption made before we conduct our empirical methods. So why can't we question it to understand its impact on our discipline? Why can't we contrast it with theism to highlight its features and its qualities?

I guess I see the answer to these questions as relevant and significant regardless of whether one is a theist. Yet I've been surprised at the ad hominem directed at me, as if the only reason I might explore the relations between these two dominant worldviews of Western culture is my own presumed dogmatism. It's only marginally accepted to discuss issues related to transcendence in psychology, but to discuss God as a centerpiece of theism is to offend any number of modernist sensibilities, especially in the academy (Slife and Reber, 2009). I've now written the lead article for any number of special journal issues on theism, and the replies to these types of articles are fascinatingly all over the map, from "you're absolutely crazy" to "you have incredible courage." Another way to put this emotional bias is that naturalism is incredibly embedded in our academic psyche—it's almost an academic sacred cow, at least in psychology.

I should probably comment a bit on the notion of methodological naturalism, to which many psychologists would say they subscribe. Methodological naturalism is typically understood as qualitatively different from metaphysical naturalism, as if the epistemology of naturalism (its methods) can somehow exist apart from a set of assumptions about the ontology of naturalism (its metaphysics). As you might guess, from my comments a few minutes ago, I don't see how these categories of philosophy—epistemology and ontology, in this case—can be completely separated. Why would someone argue for the use of methodological naturalism unless they had a worldview—a set of assumptions about the reality of the world—in which they thought this methodology would be successful? Methodological naturalism would at least be less effective in a truly theistic world, a world in which God's actions were at least a necessary condition for all events and things. In a theistic world, methods would presumably be more effective when preceded by prayer, and any explanations that didn't include God in some manner would be incomplete or flat wrong.

The real scandal, from my perspective, is the psychology of religion (Slife and Reber, 2012). Here, with few exceptions, researchers investigate the phenomena of religion using methods and explanations that cannot recognize the possibility of divine action. In many cases, these researchers are attempting to capture the meaning of prayer or ritual or forgiveness, but they are prevented from making the most pivotal assumption of the theists they're investigating—that God is vitally involved in all these activities and practices. Needless to say, they capture very little of these meanings. Perhaps worse, they turn these practices into naturalistic meanings that would be anathema to most theists. Forgiveness and prayer, for example, are turned into instrumental concepts that they advocate should be used for greater happiness and flourishing (Slife and Reber, 2012). In fact, one of the most striking differences between a naturalist and many theists is the whole notion of human flourishing. For most theists, it's not about them at all, including their own flourishing; it's about their God flourishing, whether or not human theists are flourishing in any conventional sense.

I'm also quite struck by how the reification of naturalism spawns bad scholarship (Slife and Reber, 2009). Clinicians have long admired a number of noted theists, such as Martin Buber, and have even adapted their ideas to their therapeutic work. The intriguing thing, when considering the bias of naturalism, is how all of these adaptations immediately exclude divine action in the process. Buber (1971), for example, is abundantly clear (especially in his book *I-Thou*) and in all his work that God cannot be excluded from I-Thou relationships, yet therapists who have written about these relationships have not only excluded this divine action, they have done so without any defense or justification, as if this exclusion needs no defense, as if everyone understands it can't be included, despite what the author of the ideas himself thinks. My point is that the assumptions of naturalism are as deep and as taken-for-granted as any set of assumptions in the academy. It's an amazing thing to me.

BWB: In light of your own thinking, and being a practicing clinician, how does this theism inform how you conceptualize the therapeutic experience?

BDS: I like that you refer to the conceptualization of experience, because there is undoubtedly a number of theistic psychotherapists who use their theism at least informally. They surely wouldn't reject insights about a therapy case that they might consider spiritual promptings. If they're serious theists, they'd surely pray for their clients, if not expressly ask for divine guidance. So there's likely to be a lot of informal theism and therapy going on. Your question is intriguing because it raises the possibility of whether theism can be formalized as a meta-theory from which thoughtful conceptualizations of clients and therapeutic strategies can be born. My answer to this possibility is "of course." If I, as a therapist, believe in the import of God in the world, why wouldn't I attempt to formalize this import into therapy conceptualizations and strategies (Slife, 2004b)? Allen Bergin and Scott Richards have done just that in their recent books.

I realize that some psychologists would object to the therapist's use of theism on clients who aren't themselves theistic, but this objection is funny to me—because I don't see them objecting to the use of naturalism on the vast majority of our clients who are theists (Slife, Scott, and MacDonald, 2016). Again, they tend to view naturalism as less value-laden or less assumptive, when my own investigations show naturalism to be just as value-laden and assumptive, and naturalists can easily rival theists in their dogmatism about their respective worldview. I'd also contend that theistic therapists do not have to use religious rhetoric with nonreligious clients. They can be just as spiritually prompted, and thus just as theistic, without mentioning it to the client. They can be just as guided by, for example, a Judeo-Christian ethic without mentioning it to the non-theistic client.

Truth be known, most non-theistic therapists are ultimately guided by a Judeo-Christian ethic, usually without knowing they are and typically without giving the underlying theism of this ethic any acknowledgement.

My own theistic therapeutic approach would pull a great deal from strong relationality (Slife, Scott, and MacDonald, 2016). Strong relationality is deeply embedded in the Trinitarian tradition of Christianity, for example. Colin Gunton is a great source for these ideas, as are the Patristics, the early church fathers. I have adapted these ideas into some important strategies about how virtuous relationships might occur, not to mention how to love and be loved. My own clinical practice consists primarily of people with problematic marriages, so my focus on relationships fits nicely. As you may know, Brian, a very successful therapeutic boarding school in West Virginia is based on my ideas—the Greenbrier Academy. It's a 55-bed facility for adolescent women. Even its accredited high school is based on strong relationality (Slife, 2012).

BWB: With the time we have left, I want to just wrap up with a question about how you perceive your work has been received by the larger psychological community. If we look at the field of psychology today, there has been a lot of great work done in theoretical psychology, yours being one of the principle voices in this conversation, to help psychology reformulate its assumptions in order to become less dualistic, more relational, more accepting of religion and theology. But upon looking at where the field currently is and the directions it seems to be heading, its focus on evidenced based treatments, grounding personhood in neurobiology, etc, the impact of theoretical psychology on the larger discipline is not apparent to me. How have you perceived your work being received in the discipline in particular?

BDS: It's an interesting question that I don't really consider very often. I guess I just do what seems right and good for the discipline and for humanity, and then let the chips fall where they may. I guess I care about the outcome to some degree, but I don't control it and don't fret over it. You're right, of course, that not every department is clamoring for a theoretical psychologist—perhaps an understatement. You're also right that much of the rising

> neurobiological approaches assume a problematic reductive materialism. Still, I see all sorts of hopeful signs. It probably depends on where you look. It probably also depends on how you view change or impact. When theoretical psychology is so broad brush—examining ideas that undergird the entire discipline—it might seem to follow that any impact would itself imply a similarly broad brush, but I just don't think it has to be gauged in that manner (Slife, 2000). In fact, I think that broad-brush impact is often subtle rather than sweeping, with little sprigs of their manifestations sprouting here and there.

One such sprig could be the boarding school I was privileged to help found. It surely says something about our present era in psychology that a therapeutic boarding school can be based on "ontological relationality." Ask anyone at Greenbrier what the academy is based on, and they will not only give you that phrase, but they will also tell you what it means. Moreover, there are reliable reports that Greenbrier, and especially its success, is changing the therapeutic boarding school industry. Suddenly, everyone is focusing on relationships. My entrepreneurial friend is, as we speak, putting our philosophy into modules that he thinks can be exported to the Middle East, where he believes there is a crying need for such boarding schools.

Consider as another sprig the changes we've seen in psychology's methodology. Who'd have thought even 20 years ago that psychology would be this methodologically pluralistic? I don't doubt for a minute that our theoretical writings on philosophy of social science have played a pivotal role in that important impact (Slife and Williams, 1997). You mention evidence-based practice as a negative example, partly perhaps because I've been critical of that movement. Still, any thoughtful review of those developments in psychology would see it as a move from the narrowly and naively empirical to the more broadly empirical in the spirit of William James. Is this movement where I would want it to be? Of course not. But has theoretical psychology helped it to change in directions that appear to be more thoughtful and more inclusive? I think it has.

Another sprig could be the movement toward theism we discussed earlier (Slife, Reber, and Lefevor, 2012b). Why would a half dozen leading journals do a special issue on this controversial topic if it weren't being considered seriously? And why is it being considered seriously? I can attest that it's not due to a bunch of religious fanatics trying to convert a "godless psychology." It is, instead, a bunch theoretical psychologists who are attempting to bring thoughtful and incisive scholarship to bear on important issues. Again, is this movement where I'd like it to be? No. But I fear that we often don't appreciate the subtle but real changes that occur because we increase our expectations as we go.

I can think of a dozen other sprigs, from our enhanced awareness of psychology's replication problems to thoughtful discussions about neurophilosophy to greater sophistication about human agency—all pivotally driven, I believe, by theoretical psychologists (Slife, Jennings, and Johnson, 2015). As a more personal example, my son, Nathan, recently won a national award for a dissertation that is an obvious theoretical contribution. In it, he shows how his own discipline of student affairs, specifically the literature used to train its doctoral students, is completely overrun with a liberal individualist philosophy. He makes wonderful points about how the people who are managing our college students outside of the classroom are crypto-missionaries for this ideology. I believe it's quite clear that he's helping an entire field to be more reflexive, more critical of itself in healthily skeptical ways.

BWB: So, if you were to speak prophetically right now to psychology, what would you say? What would be the message you would want to convey about what the field needs right now?

BDS: Prophet I am not, but if psychology could just be aware of its own worldview—for example, that our methods are not indisputable aspects of logic but are instead facets of an indigenous psychology—I would be in disciplinary bliss (Slife, O'Grady, and Posits, 2017). This awareness might allow us to see that our investigations and their results are works of interpretation rather than works of objective description. It would also mean that there are non-empirical aspects of psychology that need to be taken into account—theoretical aspects in which theoretical psychologists might be important resources.

I would also advocate that psychologists be trained in truly critical thinking skills (Slife, Reber, and Richardson, 2005). I say "truly" because so much of what passes for critical thinking in psychology is really just the application of logical rigor, such as the application of the logic of science. This form of thinking, I would contend, is better understood as rigorous thinking, which is not unimportant but isn't critical thinking. Truly critical thinking, as most educators know, is the knowledge of one's current assumptions as well as the awareness of possible alternative assumptions. In fact, I'd argue that we won't have a meaningful awareness of our current assumptions without an awareness of assumptive alternatives. This kind of critical thinking might allow us to always keep the "other" in view, even as we explore our conventional ideas.

BWB: So one last question, and then I want to give the audience an opportunity to ask some questions. You have a new book coming out through APA books. In that book, as well as other work you're planning, where is your own research taking you in the future?

BDS: I would certainly want to continue my probe of strong relationality and a theistic approach to research and practice. However, this exploration would necessarily require an exploration of the "others" of these philosophies, namely abstractionism and naturalism respectively. In fact, I'm sometimes more enamored with the others, because they are such fascinating hidden influences in so many fields. As I mentioned earlier about naturalism, it's the hidden system of assumptions across all the varied disciplines of the academy. Wouldn't it be fruitful to understand the impact of these assumptions as well as what a system of alternatives might provide?

I'm also interested in liberal individualism and virtue ethics (Slife, Scott, and MacDonald, 2016). The new book you mentioned is a contribution to virtue ethics, particularly overlooked aspects of humanity such as our frailty, dependence, suffering, and even evil. It's co-authored with Blaine Fowers and Frank Richardson. Hermeneutics will remain my primary inspiration. I'm currently investigating the significance of a hermeneutic moral realism, important to such scholars as Heidegger and Taylor. Is it possible to frame a nondualist, meaning-oriented approach to morality that is real and not totally dependent on our subjectivity and cultural conventions (Slife and Christensen, 2013)? Such a question may seem highly theoretical, but I view it as central to many of my clients who seem bereft of a moral compass in their practical lives (Slife and Yanchar, in press).

Bibliography

Brinkmann, S. (2011). *Psychology as a moral science: perspectives in normativity.* London, UK: Springer Press

Buber, M. (1971). *I and thou.* (W. Kaufman Trans.). New York: Touchstone Press.

Fowers, B., Richardson, F., & Slife, B. (2017). *Frailty, suffering, and vice: Flourishing in the face of human limitations.* Washington, DC: APA Books.

Gadamer, H. G. (1989). *Truth and method* (2nd rev. ed.). New York, NY: Crossroads. (Original work published 1960).

Heidegger, M. (1967). *Being and time* (J. Stambaugh, Trans.). Albany: State University of New York Press. (Original work published 1927).

Levinas, E. (1969). *Totality and infinity.* Pittsburgh, PA: Duquesne University Press. (Original work published 1961).

Levinas, E. (1989). Is ontology fundamental? *Philosophy Today, 33*(2), 121–129.

Levinas, E. (1998). *Otherwise than being or beyond essence.* (A. Lingus Trans.) Pittsburgh, PA: Duquesne University Press.

Rychlak, J. (1981). *Introduction to personality: A theory construction approach 2nd edition.* New York: Houghton Mifflin.

Slife, B. D. (2000). The practice of theoretical psychology. *Journal of Theoretical and Philosophical Psychology, 20*(2), 97–115.

Slife, B. D. (2004a). Taking practice seriously: Toward a relational ontology. *Journal of Theoretical and Philosophical Psychology, 24*(2), 157–178.

Slife, B. D. (2004b). Theoretical challenges to therapy practice and research: The constraint of naturalism. In M. Lambert (Ed.), *Handbook of psychotherapy and behavior change* (pp. 44–83). New York, NY: Wiley.

Slife, B. D. (2012). Virtue ethics in practice: The Greenbrier Academy. *Journal of Theoretical and Philosophical Psychology, 32,* 35–42.

Slife, B. (2013). *Taking Sides: Clashing views on psychological issues*. Columbus, OH: McGraw-Hill Education.

Slife, B. D., & Christensen, T. (2013). Hermeneutic realism: Toward a truly meaningful psychology. *Review of General Psychology, 17*(2), 230–236.

Slife, B. D., & Reber, J. S. (2009). Is there a pervasive implicit bias against theism in psychology?. *Journal of Theoretical and Philosophical Psychology, 29*(2), 63–79.

Slife, B. D., & Reber, J. S. (2012). Conceptualizing religious practices in psychological research: Problems and prospects. *Pastoral Psychology, 61*(5), 735–746.

Slife, B. D., & Richardson, F. C. (2011). Is Gergen's *Relational Being* relational enough?. *Journal of Constructivist Psychology, 24*, 304–313.

Slife, B. D., & Wiggins, B. (2009). Taking relationship seriously in psychotherapy: Radical relationality. *Journal of Contemporary Psychotherapy, 39*(1), 17–24.

Slife, B. D., & Williams, R. N. (1995). *What's behind the research? Discovering hidden assumptions in the behavioral sciences*. Thousand Oaks, CA: Sage Publications.

Slife, B. D., & Williams, R. N. (1997). Toward a theoretical psychology: Should a subdiscipline be formally recognized?. *American Psychologist, 52*, 117–129.

Slife, B., Williams, R., & Barlow, S. (2001). *Critical issues in psychotherapy: Translating new ideas into practice*. Thousand Oaks, CA: Sage Press.

Slife, B. D., Reber, J., & Richardson, F. (2005). *Critical thinking about psychology: Hidden assumptions and plausible alternatives*. Washington, DC: American Psychological Association Press.

Slife, B. D., Reber, J. S., & Faulconer, J. E. (2012a). Implicit ontological reasoning: Problems of dualism in psychological science. In R. Proctor & J. Capaldi (Eds.), *Psychology of science: Implicit and explicit reasoning* (pp. 459–478). New York, NY: Oxford University Press.

Slife, B. D., Reber, J. S., & Lefevor, G. T. (2012b). When God truly matters: A theistic approach to psychology. *Research in the Social Scientific Study of Religion, 23*, 213–237.

Slife, B. D., Johnson, C., & Jennings, A. (2015). Surprise and practical rationality: Knowledge advancement through the explication of interpretation. In *The Wiley handbook of theoretical and philosophical psychology* (pp. 392–408). New York, NY: Wiley.

Slife, B. D., Scott, L., & McDonald, A. (2016). The clash of liberal individualism and theism in psychotherapy: A case illustration. *Open Theology, 2*(1), 595–604.

Slife, B., O'Grady, K. & Posits, R. (2017). *The Hidden Worldviews of Psychology's Theory, Research and Practice.* New York: Routledge

Smith, J. (2014). *How (not) to be secular, reading Charles Taylor.* Cambridge, UK: Eerdmans Publishing.

Taylor, C. (1989). *Sources of the self: The making of modern identity.* Cambridge, UK: Cambridge University Press.

Walzer, M. (1994). *Thick and thin: Moral argument at home and abroad.* Notre Dame Indiana: University of Notre Dame Press.

Brian Becker is Associate Professor of Neuropsychology and Assistant Chair of Psychology at Lesley University. He is a Research Fellow at the Psychology and the Other Institute and co-chair of its biannual conference. His research explores the implications of recent French phenomenology for psychological theory and practice. He is co-editor of *In the Wake of Trauma: Psychology and Philosophy for the Suffering Other* (Duquesne University Press, 2016), and has most recently published in the "Journal of Theoretical & Philosophical Psychology and Humanistic Psychologist." Dr. Becker is a licensed clinical neuropsychologist.

Madness, Modernism, and Interpretation: A Conversation with Louis Sass

(Interviewed by Mark Freeman, College of the Holy Cross)

Louis Sass obtained his B.A. (English literature) at Harvard University in 1970 and his Ph.D. (psychology) at the University of California Berkeley in 1979. He then completed an internship in clinical psychology in the psychiatry department of Cornell University Medical Center in 1981. Around the same time, he landed at the College of the Holy Cross, where I (Mark Freeman) currently teach. It is the same position, basically, because (presumably) they would only allow a certain number of theoretical philosophical-psychology scholars to work in that department.

Eventually, Louis Sass made his way to Rutgers University. He's been a visiting professor at Leiden University in The Netherlands, the University of Chicago, *Universidad Michoacana de San Nicolás de Hidalgo* in Morelia, Mexico, the Center for Subjectivity Research at the University of Copenhagen, the Institute for History and Philosophy

L. Sass (✉)
Rutgers University, New Brunswick, NJ, USA
e-mail: lsass@gsapp.rutgers.edu

H. Macdonald et al. (eds.), *Dialogues at the Edge of American Psychological Discourse*, Palgrave Studies in the Theory and History of Psychology, DOI 10.1057/978-1-137-59096-1_3

of Science and Technology in Paris, and also the philosophy department of the National University of Colombia in Bogotá.

Louis has collaborated on research and writing projects with folks in Denmark, Spain, France, Germany, England, Mexico, Australia, Italy, Colombia, and, of course, the United States. He has been a featured lecturer in these countries and elsewhere. His work has also been featured in a number of films on the subject of schizophrenia.

From 1998 to 1999, Louis Sass was President of the Division of Psychology and the Arts of the American Psychological Association and subsequently (2006–2007) was President of the Division of Theoretical and Philosophical Psychology. In 2010, he was awarded the Joseph Gittler Award from the American Psychological Foundation, "in recognition of his longstanding commitment to using philosophy to advance psychology research and scholarship."

Among his many writings are the books, *Hermeneutics and Psychological Theory,* which he co-edited with Messer and Woolfolk; *Madness and Modernism: Insanity in the Light of Modern Art, Literature and Thought (which will appear in a revised edition in 2017)*; and *The Paradoxes of Delusion: Wittgenstein, Schreber and the Schizophrenic Mind.*

It's an extraordinary profile. We are very fortunate to be able to have this conversation with him, and to top it off, Louis Sass is a good man with a great mind. It's actually been a long time since I've seen him. He's gone his way and I've gone mine, and we just haven't had the opportunity in recent years to meet up with one another to share ideas and to find out what's been happening. But Louis is someone whose questions and ideas and ways of encountering the world are so familiar to me that, even after a too-long lapse of time, we can connect. It's quite remarkable, actually, to find this sort of intellectual kinship. Among other reasons, it helps one see that, through it all, we can sometimes look out on a common world. Either that or we partake of similar illusions! One way or the other, it's my great pleasure to introduce Louis Sass.

Mark Freeman (MF): I want to begin today with a quote. It's from a review of the book I just mentioned, *Madness and Modernism.*

> *Madness and Modernism* possesses extraordinary richness, depth and profundity. It is truly a work of thought, a work that not only explodes received wisdoms, but also allows its readers to see the world anew. In this respect, the book itself is a prime example of one of the fundamental theses it seeks to articulate. To become ensnared in the labyrinth of our own constructions is to become blind to the otherness of the world. It's to become autistically enclosed, hermetically sealed, such that everything appears to be quite other than what it is. More than anything, therefore, Sass is urging us to see, to look again in the hope that we might learn something new about our situation.

As you know, those words are mine from a review of the book. I had to look up the review; that was about 20 years ago from the journal *Theory and Psychology* (1996). And I must say, they reminded me of how extraordinary your work has been. That book, I think, is a classic in the field, but even beyond the field. It's just an amazing work of thought. I mean, even the endnotes, 160 pages of them! Very, very intimidating—but a great book.

In any event, those words seem especially pertinent to the conference here today: the relationship to otherness. What happens when that otherness becomes eclipsed? What happens when one turns inward in such a way that the world becomes veiled? So, I'm going to eventually ask you to talk about it.

It was almost 30 years ago in conjunction with a conference on development in the arts, run by our friend and colleague, Bernie Kaplan, that we first met. And it was there, I think, that we felt an affinity. We felt that we were fellow travelers of a sort, especially when it came to thinking about the assets and liabilities, so to speak, of postmodernism in the arts and beyond. How did it all begin? I realize that's a very broad question. How did you enter the discipline and what did you make of it at the time? If you could just kind of sketch out a narrative of those formative years and help us see how you got to be you.

Louis Sass (LS): I'll do that in just a second. But first, I want to thank you, Mark, for being here and doing all this preparation when you have so much else to do, including being an interviewee yourself at this conference. Thanks also to Heather and David for inviting me, and to all of you for coming. It's really a pleasure to be here talking with you.

So you asked me how did I get into the field?

MF: How would you get into the field? Or what led you to the discipline of psychology?

LS: Well, it was a sort of strange journey in a sense because I started out as an undergraduate majoring in literature, English literature. And only took a couple of psychology courses and then wandered into psychology, for graduate school, having very little sense of what the field really involved. Wandered into it because of several things I had read as an undergraduate and the couple of courses I did take; two books were particularly important: David Shapiro's *Neurotic Styles* (1965), one of the classics of our field, clinical psychology; and also R. D. Laing's *The Divided Self* (1960), which also impressed me very much.

So I . . . I wasn't very career oriented, not in any focused way, and thought, well, maybe there would be more options in psychology than going on to graduate school in literature. And so it was mainly on that basis. I went to Berkeley because I wanted to be somewhere different (I had majored in English at Harvard College) and I figured California would be an exciting place to live for a while. . . . not knowing much about who was on the psychology faculty there, or anything serious of that nature. So it was really all a bit random—I certainly wasn't a savvy 21 year old.

But I had gotten interested in certain things like schizophrenia because of Laing, among other reasons, and also in phenomenology because, almost by accident, I ended up in a philosophy course on Husserl and Merleau-Ponty toward the end of my college career. And I can say that I did read every word of *Ideas 1* by Husserl, though I think I probably understood almost nothing of it. But I did read it. It's an interesting question, what does it mean to read? Because I, in some sense of the word, did read that book, that is, passed my eyes over every sentence, even though I didn't understand anything really. But I also read Merleau-Ponty's *Phenomenology of Perception*, which I think I understood some—maybe quite a bit of.

MF: Who was the teacher, I'm curious, do you remember?

LS: Yes. It was Frederick Olafson. He left Harvard about a year later, for the philosophy department at the University of California San Diego. He wrote a lot on phenomenology, mostly on Heidegger, I believe.

I found—there's a lot of psychology in Merleau-Ponty's *Phenomenology of Perception* (1945/2012), and I thought, "Wow, this is really interesting stuff." So I expected to arrive in Berkeley, where I went to graduate school, with everyone in the psychology department knowing all about Merleau-Ponty and being eager to talk about these things. And of course it wasn't at all like that. No one seemed to have even heard of Merleau-Ponty; and hardly anyone, at least in psychology, was interested in talking about the things that interested me. So that's what happened.

Then in graduate school, I had a difficult time in some ways. I was always thinking about dropping out, and I did drop out a couple of times because I really didn't know that the field was for me. Much of the time I thought there was something very wrong with the field. But there were many other times when I thought maybe it was me: maybe I was just too stupid to appreciate the subtleties of something like mainstream theories in social psychology, a subfield that particularly irritated me. So I certainly had moments when I thought, "Well, I must just be missing something, I don't understand it. I just don't understand what people see in this to take it so seriously."

It was a difficult time. But fortunately, I met Margaret Singer, a clinical psychologist who studied schizophrenia and then I managed to find some direction and a dissertation topic. So that's kind of how it went for me.

MF: I also sort of stumbled into phenomenology by accident. It was a course in phenomenological psychology in the philosophy department. I really had no idea what that meant. But I knew that I wasn't getting fed, so to speak, in the way that I wanted to from the courses I was taking. And in encountering phenomenology . . . I had a couple of thoughts. One of them was I thought this was what psychology actually is in its essence, or should be anyway. So, it was a kind of natural—a natural thing. Did you feel the same way? Did you find a kind of home there in a way?

LS: Absolutely. Because, well, I think psychology should be largely about human subjectivity—including of course the relational or intersubjective and the bodily-based or corporeal aspects of human subjectivity. Those are certainly part of it. And there are always other possibilities as well. But the main thing that should be at the core of psychology, in my view, is human consciousness, human subjectivity. And, of course, phenomenology—with all its complexities, with all its internal divisions and dissensions—is the royal road to that.

MF: Sure. It's amazing how many different iterations there have been in the history of psychology of avoiding encountering subjectivity, from the behaviorists on up, in some ways, to contemporary neuroscience. I mean, there are some neuroscientists who are interested in subjectivity, but oftentimes they're more interested in the material foundations of it as if we could safely sidestep subjectivity.

LS: Absolutely, yes.

One more thing about that period: I think getting interested in schizophrenia was a very natural thing for that period of time. You know: the late' 60s, everyone speaks of "sex, drugs and rock-and-roll." But actually, it was sex, drugs, rock-and-roll—*and madness*, in terms of the themes that really were passionately interesting to people of my generation, in that phase of our lives at least. And these were, well, the cool things—so to be interested in schizophrenia was a very natural thing; and it was kind of, in the late' 60s and into the' 70s, it was sort of *'the'* disorder.

MF: Right.

LS: Yes, in a way that is not really true now, either in the mental health professions or in the culture at large. It's still important, obviously, but it's not "*the*" thing. So in that sense, I suppose I was, in a way, very much a product of the passions and enthusiasms of my era.

MF: I understand that. I mean, rereading R. D. Laing's *The Divided Self*, which I've done in recent years, it's kind of a trip. It's an amazing, amazing book and I think it's been set aside in many ways. But you look at those chapters on ontological anxiety and the false-self system and so on, they're amazing. I think of some phrases from *The Politics of Experience* when Laing, for instance, talks about socialization as "the indoctrination of recruits"; those are still important.

LS: Yes, I agree.

MF: And a lot of that was about liberation in a way.

LS: True. I was influenced by *The Divided Self*, though.

MF: Same here.

LS: And I was—I have always been very dubious about almost everything else written by Laing. Although it's interesting—he was, as you know, often brilliant—but I would make a sharp distinction between *The Divided Self*, which is in the phenomenological tradition, and other things written after Laing began to think of himself as a guru with all that that implied, and when lots of problems began to happen in his life and with his work.

MF: Quick comment on *The Divided Self*: I've had a number of students who wanted to study things like eating disorders through the years. And one of the first books I had them read was *The Divided Self* even though there's no explicit mention of that.

LS: Yes.

MF: The whole idea of what one puts out, so to speak, for consumption and what gets negated and impoverished internally and so on, is spelled out in quite extraordinary detail in that book.

LS: Yes, I agree.

MF: A number of years after we met, we gathered again at a symposium at the APA convention. The topic was "Post-Modernism and Its Malcontents"; I mentioned it yesterday. We were among the malcontents and our friend, Ken Gergen, was our discussant. I'm actually not sure whether "malcontent" was quite right. By all indications, you were, and remain, attracted to certain aspects of post-modern and post-structuralist thinking. You've talked about how it is that our own self-understanding might be enriched by at least certain aspects of schizophrenic experience. But you clearly had misgivings, too, about what you were seeing in terms of intellectual currents in psychology. What were the problems that you had? What is it that concerns you most about those bodies of thought? Why is it that you became a skeptic of the skeptics?

LS: Right. Well, as you well know, Mark, as well as anybody does, words like post-modernism have lots of different meanings.

MF: Sure.

LS: As does modernism, by the way. But post-modernism, if by that you mean a kind of relativistic position, a dogmatic relativism, which denies the project of seeking truth, then I'm certainly *anti-* post-modernist because that just seems to me a naïve position, and one that is basically "metaphysical" in the pejorative Wittgensteinian sense of insisting on some grand generalization (in this case, a skeptical one) that cannot be justified either by science or by common sense.

So, in that sense I'm *anti*-post-modernist. I mean, as I said, it seems to me that, as an epistemological position it goes against common sense. It goes against the fact that you and I are speaking with each other now, and that this fact presupposes that we, each of us, believes and is necessarily committed to the idea that each can indeed have some idea of what the other means. With all sorts of possibilities of error, obviously.

MF: Right, and ambiguities.

LS: And ambiguities, certainly. . . . But if there weren't some possibility of understanding, and in that sense of seeking truth—like I can try to figure out: what does Mark mean by his last question?—then there would be no possibility of our even talking. And if there's no possibility of talking, there's no possibility of thinking. So I find that position doesn't make sense to me. It seems to me it's more a kind of posturing than a serious intellectual stance. And, of course, there are also the familiar arguments against skepticism, for example regarding the self-contradictory nature of absolute skepticism itself. So no, I don't find that version of postmodernism, at least as a general epistemological stance, to be at all appealing, or even reasonable.

But of course, the rubric "post-modernism" can include a lot of things. If by that you mean to include what's called post-structuralism, people like Michel Foucault, for example, well obviously then, there's a lot of interesting stuff there. Foucault, to my mind, is perhaps the most important thinker of the last 50 years or so. But I do think that there's a strong tendency to think very simplistically about who fits with whom. And part of that is fostered by people like Foucault himself—who, after an early phenomenological period, *attacked* phenomenology, and allowed himself to be seen . . . indeed supported the idea that there was a radical separation between the sort of work he did and what phenomenology was all about.

Which I think is completely misleading. I've written about this recently, discussing both Foucault and Lacan in this light. Despite the fact that both of them are overtly anti-phenomenological, judged by what they say explicitly, I do think an important way of reading both of them is as, in a way, secret phenomenologists—of a specifically Heideggerian sort (Sass 2014a, 2015).

MF: Yes. Why do you suppose they had to keep it secret? In other words, why the disavowal?

LS: Parisian politics . . . Parisian intellectual politics, to a large extent, and also the need to, well I guess you could call it a kind of anxiety of influence: the need to present oneself—to others but to oneself as well—as radically different, truly novel, even revolutionary. Also, the fact that they were, indeed, influenced by Claude Levi-Strauss, whose own position was in fact more truly anti-phenomenological.

MF: Right, for sure.

LS: And Levi-Strauss presented that in very sharp terms, for example in his book *The Savage Mind.* So they got on that bandwagon—and then there were also all the Americans who were influenced by the Parisians and followed (perhaps a bit slavishly) that same line, assuming that that must be the right way to look at things. In fact, if you read almost all of the major histories of recent French thought, which I reviewed recently, they all assume that there was a major bifurcation—between existentialism and phenomenology versus the poststructuralism that followed. But I don't think that really makes a great deal of sense, not, at least, once you recognize or accept the crucial role of Heidegger. Because Heidegger is obviously a central figure of the phenomenological tradition. And yet, Heidegger is also, I think you'd have to say, a very major influence on both—on Foucault (along with Nietzsche) and on Lacan (along with Freud).

So, when you think about it, the standard historical narrative doesn't make an awful lot of sense. And then you begin to think about, well, what is Heidegger really all about and in what direction did Heidegger take phenomenology? And the answer, I think, is that the direction he took phenomenology (beginning at least with *Being and Time*) is very consistent with a lot of what both Foucault and Lacan have to say. Indeed, we should not underestimate the extent to which they are influenced by Heidegger's brand of phenomenology—especially by his crucial emphasis on not "forgetting the ontological difference," which shows up, I think, in key notions like Foucault's "epistemes," in his book *The Order of Things* (originally: *Les mots et les choses*), or Lacan's three registers (Imaginary, Symbolic, Real). (On Heidegger's notion of ontology and the ontological difference, see Sass 1992a, 1992b, 2017)[1]

[1] Heidegger (1927/1996) describes the "ontological difference" (p. 72) as the difference between "being and beings." He directs the phenomenologist's attention not to any particular object or 'ontic' entity (or "being"), but to what he calls the 'theme of ontology,' which is the overall *way* in which everything shows up, especially regarding its felt quality of reality or the lack thereof. . . . It is this most general ontological dimension—call it the world's form or manner of presence (its "Being")—that is, for Heidegger, the very heart of our existence as subjective creatures yet that is

That's my argument, anyway. So I'm against *relativistic* postmodernism yet, at the same time, very interested in some of the post-structuralist postmodernists.

And one more thing I should say perhaps: which is that I'm also very sympathetic to the skeptical moves of postmodernism so long as they're used as intellectual strategies and not taken to an absolutist extreme. I mean, I do think that postmodernist psychology, in questioning the mainstream, has had many interesting things to say. By bracketing the truth-value of psychological theories, and saying instead, "let's look at these psychological positions in terms of how they reflect the culture, hidden intellectual presuppositions, and things like that." This can certainly be an interesting project—certainly it is one that I respect, and that in my view has generated a lot of important writing.

I do recognize that such a project does require skepticism, a bracketing of truth claims, or at least of many such claims, within a certain domain. So you see, there are indeed several ways in which I'm actually very sympathetic with much of the work that is considered part of postmodernism.

MF: It would be great if you could clarify one thing. I think I know where you are on this, but some people might be taken aback in some ways by the fact that you—or I, for that matter—don't subscribe in a wholehearted way to the kind of constructionist wing of postmodernism and so on. In reading some of your work, one could be led to assume that you're something of a realist, which I don't think is necessarily as pejorative a term as it's sometimes made out to be.

At the same time, of course, you self-identify as a hermeneutic thinker. This came up yesterday in the talk with Ken. I mean, how do you square, or how do you reconcile, your concern about... how shall one put it?... preserving the real, or even more to the point for this

so readily forgotten, ignored, or distorted by reification and other distortions that seem to come as naturally to us as breathing. (from Sass 2014a, p. 329, slightly altered).

conference, preserving the *otherness* of the real? How do you square that with the hermeneutic perspective—which is itself sometimes seen as a constructionism, right?

LS: Right, right. But as you yourself commented yesterday in that interview with Ken Gergen, hermeneutics need not be seen as a relativism, and I don't think that's the proper or best way of understanding, for example, Gadamer. And Gadamer comes right out of Heidegger, and clearly it's not the right way to interpret Heidegger because he's obviously making truth claims all over the place (see Messer et al. 1988; Sass 1988; Sass and Woolfolk 1988).

So, I think that's a distorted, oversimplified view of hermeneutics: to say that it's fundamentally relativistic. Of course, hermeneutics recognizes all sorts of things that are *related* to a recognition of perspectivism and ways in which—

MF: Context shapes the perspective.

LS: Or that one's position could be constructed and maybe even distorted by cultural or epistemic formations of different kinds. But that doesn't mean that it's dramatically skeptical or ultimately skeptical about truth. (We should remember, after all, that the very act of recognizing these supposedly determinative cultural or epistemic formations is *itself* fraught with truth claims—that is, with claims about the very nature or existence of these formations.) Of course we're talking in psychology, hopefully, and certainly in phenomenology, about a realism that is not concerned with something that is real in quite the same way that physical objects are real. It's about subjectivity. And what it is to study subjectivity is, of course, a fascinating and difficult problem in itself.

We obviously need to recognize the difficulty of pursuing this project, and the phenomenological tradition is full of reflections on that issue, often with one phenomenologist criticizing another such as when Heidegger criticizes Husserl, for example. So we need to recognize all of the difficulties inherent in being, in a certain sense, realists about the, in a way, very *unreal* real thing that is subjectivity.

MF: Yes. And in one of the pieces that you sent along, which I very much appreciated, you talked about, in a sense, the possibility of taking something like an objective stance toward subjectivity.

LS: Right.

MF: Which I'm very sympathetic to. Say a little bit more, though, about what that means, because on the face of it, it sounds kind of paradoxical. How can one be objective about the subjective? But it seems to me that in some measure, that's what phenomenology's about and that's certainly what much of your work is about.

LS: Here I am reminded of the quotation from Eugene Minkowski (the French phenomenological psychiatrist who published *La Schizophrénie* in 1927) (Sass 2001b) that appears as a motto at the beginning of *The Divided Self*, which you may or may not remember. Laing quotes it in French, as I recall, but basically it says something like the following: "I am engaged in a subjective project here, but one that nevertheless strives, with all its powers, toward objectivity." I think that's a beautiful statement—like so much of Eugene Minkowski, who has been a great inspiration to me.

So that, I would say, is my project—and indeed it is the project of most of us who are phenomenologists. That is, we are trying as best we can to get it right—like in my case about schizophrenia. We're not claiming we *have* it right, of course; but nevertheless that's the goal we're aiming for. Obviously, our project is not just to make up interesting things about this or that illness that might be fun to think about. The point, of course, is to reflect, as best we can, what it really is, which is to say, what it really is *like* to be this or that kind of person. And I think every phenomenological psychopathologist would agree with that. That's our project; and that's why we consider it a serious project—and one that is, in part, a *scientific* project.

But it's a difficult project especially if you think about something like schizophrenia because, well, if you consider what are the criteria of truth in understanding subjectivity, one criterion is—it's a hermeneutic criterion—what is known as the principle of charity in interpretation. And the principle of charity in interpretation, as you well know, states that, other things being equal—which they aren't always—but other things *being* equal, the construal of what the other person is saying that ascribes to him or her the greatest rationality and coherence, is probably the best one.

We rely on that all the time—even in everyday conversation. And certainly we need to use that when we're attempting to understand, say, psychiatric patients. But what happens when you're dealing with something like schizophrenia—where our notions of coherence (and of rationality, for that matter), those of the average or normal person, might be different, first of all, from theirs? Also, there may be certain things in their point of view that are actually *not* coherent, that may *be* paradoxical.

So, that means that a straightforward attempt to make the patient make as much sense as possible—at least in *our own terms*—can very easily be misleading. And so, in order to carry out the hermeneutic phenomenological project properly, one has to be extremely self critical and self aware of what one is doing in the act of interpreting. And that's what I've tried to do in, for example, postulating certain kinds of paradoxes in schizophrenia—because to do this is to acknowledge forms of incoherence that may nevertheless make a kind of sense—paradoxical though it may be. (See, e.g., Sass 1992b, 1994a, 2004a, 2007, 2014d, 2017; Sass and Byrom 2015b.) So, my project is not simply to say, straightforwardly, that whatever the patient says makes perfect sense—full stop. What I am saying is, rather, that it may make a certain kind of sense *within* a strange world, the horizons or dimensions of which have been altered from the normal, and which may well involve forms of self-contradiction.

MF: Yes, I see what you are saying.

LS: But when do you know—just one more point—when do you know to apply the principle of charity in a straightforward way, versus when to look rather for something paradoxical? Well, obviously there's no rule for that. You try things out, you try to figure out what interpretation fits the phenomenon best. And one of the things that I take into account in doing that—it's also one of the reasons there were so many pages of endnotes in my book *Madness and Modernism* (1992b/2017)—is the empirical research that is available, insofar as that is relevant. Just *how* it can be relevant to a phenomenological project is a complicated question. But often it is. For instance: what do we know from the empirical literature about the specific nature of the formal thought disorder that is most characteristic of schizophrenia—as opposed to forms common, say, also in mania?

And ultimately, what do we know about neurobiological alterations that might or might not be consistent with a phenomenological account—or that might suggest a different one? So, what I advocate is certainly not phenomenology to the *exclusion* of these other approaches.

MF: I must say, your willingness to make significant contact with the empirical work, including the neurobiological, is something I very much admire. Quite honestly, I don't do it as much in my own work as you do and I think you're a model for how to have a more inclusive sense of things.

So let me ask you one difficult question having to do with much of what you just said. To what extent do you feel at this point that you "know" the inner world of the schizophrenic person? Or would you have to be it to know it?

LS: Well, I—that's a really interesting question, and another complicated one. I have a friend, a very good friend, Rupert Read, who's a Wittgensteinian philosopher in England: we get along very well; I always learn a lot from him; and he's a critic of my point of view in a way that's very relevant to what you're saying. Because Rupert says that what I and other phenomenologists say about schizophrenia is all a fantasy.—We *can't* really know what it is like to experience such a form of psychosis. And, in fact, even understanding, the very *possibility* of understanding is itself limited by the common sense point of view that we have. That's his—that's Rupert's view of Wittgenstein; but it's not mine.

And so, almost by definition (according to Rupert) a normal person—whose speech and conceptual structure is necessarily grounded in normal common sense and normal usages of language—cannot truly get *outside* that structure in order to understand a different one. So Rupert—he says he admires my work because he thinks it's the *best* phenomenological interpretation available, but at the same time he also thinks that the whole project is probably a grand illusion. So as you can see, Rupert and I have an interesting conversation.

But I don't think—to me, that position just doesn't make sense. For one thing, it is contradictory. Just to give a very simple answer, or the beginning of one: I'm in contact with quite a few people with schizophrenia and quite a few of them tell me that what I and some other phenomenologists have written about the lived world of schizophrenia is largely correct. And I do always ask them what might be wrong in what I have written. And sometimes they say, "Well, you've exaggerated this, you got this a little bit wrong, or you neglected that." But overall, I felt quite vindicated by their reactions—and vindicated in my critique of mainstream views of the psychological nature of the condition. By the way, I myself have worked quite a bit on Wittgenstein (Sass 1994a/1994b, 2001a, 2001c, 2003a), and my own views about the implications of Wittgenstein's thought clearly differ from Rupert's.

MF: Good.

LS: Maybe it's not surprising that I have felt vindicated by the reactions I have had from people with schizophrenia—because, after all, it's not as if I wrote these things without previously *talking* to people with schizophrenia, and reading what some of them had written. I was already consulting with them, obviously; so what I say is partly a reflection of those conversations. And then I meet other people with schizophrenia later in my career and they also confirm it, while also sometimes pointing toward some new lines of thought about it. So I feel... well, I guess I don't think of course that any of us can *fully* understand any other person.

MF: Right, of course not.

LS: Another thing I wanted to say about all this is that, as you know, my book is titled *Madness and Modernism*. The whole argument is that, if you understand something of what modernism is, especially in its hyperreflexive and alienated aspects (this would include a fair bit of what is also sometimes called postmodernism), well then, you may be in a position to have a better understanding of many aspects of schizophrenia (Sass 1985, 1992a, 1992b, 2017). You don't have *exactly* the same thing as schizophrenia, of course. There's a great line from Coleridge about the necessary inadequacy of all comparisons: no metaphor, no likeness, goes on all fours. So it's obviously not *identical*.

But nevertheless, there are a lot of things that you—that you can at least approach, or understand partially; and especially, perhaps, if you have a little bit of a schizoid quality to your own personality—which many intellectuals do have. When I say schizoid, I don't mean it in the DSM sense, but in the sense of Fairbairn, Guntrip and Laing—the British object-relations theorists. These thinkers understand schizoid detachment not as a deficit of social interest or capacity but as an, in part, defensive reaction that goes together with forms of sensitivity and felt vulnerability—also with awareness of the solitary nature of the human condition; and finally, with a propensity to stand back from conventional frameworks or taken-for-granted assumptions, to float outside one's own stance, to see things from afar or adopt a meta-perspective. And I think that, in that sense, most really interesting philosophical minds, like Wittgenstein's, to take one clear example, do have a considerable schizoid streak (Sass 2001a). So, I guess I'd like to think I might have a little bit of a streak myself.

MF: Yes, a little hyperreflexivity comes in handy once in a while. [laughter]

LS: That's right, when it's not tripping you up.

MF: Exactly. I'm sympathetic to your response. You know, in a similar way I mentioned my mother yesterday, who's a 92-year-old woman with dementia, and I've written extensively about her. And the only challenge there is that she can't confirm for me whether what I say is valid. So that's difficult. But the idea that by degrees you might be able to, in some sense, enter—enter into that world and at least imagine what it's like seems to me to be plausible.

LS: What you say, reminds me that one of the great problems that we have, one that can't really be surmounted fully in phenomenology, in the phenomenology of schizophrenia certainly, is that we are probably more dependent on a certain kind of patient's account than we might like to be. But understanding the subjectivity of another is not, of course, *wholly* dependent on their personal reports.

MF: Of course not.

LS: Because we have other kinds of data as well, from experimental psychology, cognitive science, and neurobiological research; but still, verbal reports from the patients themselves are really quite crucial. And so you have to keep asking yourself, "Am I describing things that are only true for the most articulate patients—the ones likely to have the greatest influence on our phenomenological accounts? Or are the most articulate patients articulating what the others are experiencing but can't describe?" Having any certitude about how to answer that question is pretty unlikely. But we do our best.

MF: Yes. Let's go back to your work *Madness and Modernism*. What we have in the relationship between the two is a correlation, so to speak, right? Elements of the former and elements of the latter seem to be somehow of a piece. So how do you really understand this correlation? And why should it be? Why is it that we have large cultural trends that mimic the kinds of pathologies that we see in schizophrenia and related maladies? Are there aspects of contemporary culture that are simply "ill"? What do you make after all these years of this correlation?

LS: Well, what I would make of it now is no different from what I said in 1992 in *Madness and Modernism*. Actually, I often think that almost everything I have said since has its roots there, and is even stated in that book. (I realized this recently when I went through the entire book again to prepare a revised and updated edition—which is scheduled to be published in 2017 by Oxford University Press.) In any case, in *Madness and Modernism* I have an epilogue on schizophrenia and modern culture, at the outset of which I state that this book is *not* about causal associations—it's about structural formal affinities; but that I am going to raise the causal question nevertheless, here in this epilogue. And I'm still not sure quite what to say about those affinities except that, well, I do think there are things about modern culture that encourage hyperreflexivity. You know, one can understand that aspect of modern culture in a, let's call it, Hegelian fashion—in the sense that there may be some kind of intrinsic press toward self awareness (as well as toward an associated self-alienation) on the part of the human mind or spirit, a trend that, perhaps, reached a certain kind of culmination in Kant and later when we began really to understand the sense in which the very world in which we live is constructed by us—and that of course leads into phenomenology.

You can also understand it in a more Foucauldian way in terms of concepts like "panopticism," where the social and institutional structures of society are such that they involve the encouragement of self-consciousness on the part of the subjects in that culture as part of the "disciplinary" movement of modern society. And all that could certainly have its effect on people with schizophrenia—who are, of course, members of the society, and members who may also have a genetic vulnerability toward psychosis or something like that. And so, the difficulties take this particular form in modern culture, which may be different from the forms they typically took in pre-modern culture. (See Sass 1992b, epilogue, 1994a/1994b, 2004b, 2017, epilogue; also McGilchist 2009.)

MF: What do we know about the history of those forms?

LS: Well I go into all of that in detail in the Epilogue to *Madness and Modernism*, and there I survey and analyze what's been said about the issue. The question of whether there was schizophrenia, or much of it, before around 1800 is a contested issue. But it is at least *plausible* to say that it was much less common before 1800. I don't claim that we can say that with great confidence, though. This would get us into a lot of technical arguments, because you are looking back at descriptions of patients before the advent of modern psychiatry, and it can be difficult to know what to make of those descriptions, et cetera, et cetera. But I don't think it's at all unlikely that schizophrenia was indeed less common before around 1800—and indeed, several experts who have looked into the issue most carefully, like Richard Warner and Edward Hare—that's exactly what they do argue based on their careful examination of the historical evidence.

MF: Interesting.

LS: So, you see, the empirical data might well be consistent with the possibility that we don't see specifically *schizophrenic* forms of psychosis, or at least, not very many instances of them, until after around 1800. (This of course raises the question of just how we define schizophrenia.) So, it could be the case that things like an increase in certain forms of self-consciousness—of hyperreflexivity and alienation—on the part of members of modern culture might manifest itself in schizophrenia.

Of course it could also be that there are certain things that are more neurobiologically determined that simply *mimic* modernism. And probably it's a bit of both.

MF: Yes. It's interesting to begin to think about how it is that certain forms of cultural life can make manifest certain forms of pathology that come to acquire precisely the kind of objectivity and so forth that you've spoken of. So, let me ask a question. At the beginning of your answer to that question you mentioned a kind of Hegelian idea where, you know, maybe this has to do almost with the kind of self-realization of individual autonomy or something like that. So let's continue the dialectic for a moment. Is it possible that there will be another swing somehow away from this? In other words—and I'm not just thinking about schizophrenia here, I'm also thinking about, you know, the various theses proclaiming the end of art and so on— is it possible that at some point there will be some sort of reclamation of otherness that's been lost through this hyper-reflexive mode?

LS: Reclamation of otherness?

MF: Reclamation of otherness, that's right. Reclamation of more of a living relation to the real than what one finds in the hyperreflexivity of schizophrenia and in certain forms of modern and postmodern art and literature and so on.

LS: Yes. Perhaps so. But not being a futurologist, I'm not sure quite what to say. And I know that you've thought about this as much as anyone because of the fascinating things you wrote about the postmodernist artists, which I think should be much better known, in your book *Finding the Muse: A Sociopsychological Inquiry into the Conditions of Artistic Creativity* (1994). And the. . . .

MF: You take what you can.

LS: Right. One should consider the weird position in which people who follow what one might call the Duchampian trend find themselves, as you well know, even if they're successful in some fashion.

MF: That's right.

LS: When art is just about art is about art is about art, you know? And—

MF: And many of them eventually leave that mode. That's kind of where I'm going. In other words, that reaches a kind of dead end for some people and in some cases they'll ask themselves, Does this have anything for me or am I out?

LS: Well, you know, some of the influential younger novelists—whom I have mostly not read in any depth and don't know well, like David Foster Wallace and David Eggers—seem to have been writing about this, and criticizing the overemphasis

> on irony, and seeking something else. Probably some of the younger people here in the audience know these writers better than I do.... How much they really transcend the hyperreflexive trap: that I don't really know.

I guess I'm still fixated in my own cultural moment, with its cultural memory and its icons—which is before they wrote, and I'm more interested in that; maybe I'm stuck in my ways. Anyway, when I think of people who have a way of overcoming the bad aspects of hyperreflexivity, two people come to mind. One is Louis Armstrong: especially if you listen to Louis Armstrong singing, and especially some of the *earlier* singing by Louis Armstrong ("When you're smiling," "Jeepers creepers," "A kiss to build a dream on," to mention just a few examples—but also the sublime, erotic/ironic duos with Ella Fitzgerald, which come a bit later, I believe). Well, the combination of irony and self-mocking amusement at himself together with love of the other—love that is deeply felt and authentically expressed—is just unbelievable. I mean, what a genius—and more.... I mean, it's like—when you recognize that in him, it can be a very deep spiritual experience, I think, to recognize that. And that's a sort of quintessential example: I mean, the birth of jazz—obviously one of the greatest contributions of modernism is jazz. So there you have it: this fantastic combination—of something incredibly vital, but that can also be supremely ironic and self-aware. And I think it's no accident that it's also linked to the erotic—to the erotic at its highest and most sophisticated level: which is partly playful and self-mocking, but also deeply engaged, and tuned in to the tragic.

So I think of that as a sort of direction that one could follow. The other example I would mention is the poet Wallace Stevens, in a completely different style where you have some amazingly intellectual and self-conscious poetry, often *about* poetry—very *meta,* in fact—but that also retains so many elements of heartfelt romanticism, with all its commitment to authenticity of expression and love of the world (think, for instance of a poem like "Sunday Morning"). So I think it is possible to not let hyperreflexivity take over even though it remains present; I mean, there's just nobody more sophisticated than Louis Armstrong or Wallace Stevens, in their essential sensibility and point of view. And so it's possible to be extremely hyperreflexive, without losing touch, or connection with the other.

MF: Let's get back to your own career. You know, one of the things I've always been really amazed by in some ways is just the range and distinctiveness of your intellectual project. You range, as we know, over phenomenology and hermeneutics, but also esthetics and art history, literature, literary criticism, psychoanalysis, photography, and so on. That's hard to do. What were some of the challenges that you faced in becoming the kind of psychologist—and, more broadly, the kind of intellectual—you've become? I mean, there's not as much room for the likes of you in the academy, especially, as there might be. Somebody who is truly venturesome and willing to explore the boundaries of things and so on. What's it been like? Has it been an easy ride, tough ride, bumpy?

LS: Both, both. It's been bumpy at times *and* I've had a lot of luck in a lot of ways. But, I mean, the biggest problem, really, I mean, this is no surprise, is just being in the field of psychology, 98 percent of which—at least in *academic* psychology in this country—is not attuned to the kind of work that I do or am deeply interested in. But a lot of people in this room, I'm sure, feel that as well. Well, there were times when I thought *I* must be stupid. And there were times when I thought *the field* was stupid and I had better get myself out of it. I've gone through a lot of different thoughts like that, especially when I was younger.

But I've also had a lot of good luck. And I think one of the major sources of good luck was at a certain point when I was seriously considering getting out of psychology, I got a fellowship at the Institute for Advanced Study in Princeton, New Jersey . . . really mostly because of [the cultural anthropologist] Clifford Geertz, who was a professor there and who was, and remains, one of my intellectual heroes. And Geertz liked my proposal, which was a proposal for writing *Madness and Modernism*, which I thought I would write in about a year. Ten years later, the book comes out.

MF: Exactly.

LS: And so I also had people like that who gave me an opportunity, and then I got the job at Rutgers and my fellow department members in the Department of Clinical Psychology at Rutgers are people who, for the most part, have been open to other options—sometimes in some of their own work, sometimes in the sense of live-and-let-live. And so, even if that's not *their* interest, they have let me do my own thing. So I think a lot of people would not have that, you know—often, one doesn't have that kind of good luck.

MF: No kidding.

LS: So. There are a lot of things I could say about that, but maybe—

MF: The luck thing is key. You know, you've got to be in the right place at the right time.

LS: Yes indeed.

MF: So just a quick question about psychology. So if you had to do it over again, would you still be a psychologist? You've mentioned a number of times the temptation to flee. Is there enough in this discipline? I mean, I deal with this question with my students all the time; many of them are tempted to flee because the discipline isn't what they imagined it to be. It's often alienating and difficult, and so on. I'm able to say to them for the most part, "You know what? At its best, it's still a really cool discipline. Hang in there, it's going to be a tough road, but let's see if we can make it work." Where are you on that issue?

LS: I think the same place you are, that's just what I would say and do say. And, I mean, after all, the phenomena that we're, so to speak, licensed to talk about as psychologists, are very interesting phenomena. For example, psychopathology. And, you know, if you're interested in those phenomena, there aren't too many other options. When I was younger, I often thought that I would have been happier if I'd gone into philosophy or literature; that I would have been much happier if I were a professor in one of those two fields. But over the years, I'm not sure I think that anymore.

MF: Same here.

LS: Because I look at what—well, for example, literary studies, which—I mean, there are a lot of people I greatly admire and things being done there that I have taken from. But there are lots of problems with these fields as well. And the fact that in psychology, we're really talking about real-world phenomena, subjective though they may be—indeed, subjectivity itself—I'm very happy about that because I feel that at some level, I'm like a scientist confronting reality. What I mean by that is that I'm talking about something in the real world—not mainly about *representations* of the real world, which is what literary studies is, largely. (I realize, of course, that representations are also quite real, in their own way—but still. . . .) And philosophy is so, you know, well, there are such problems with contemporary analytic philosophy that I don't even want to go there—even though there are certain thinkers I greatly admire and have been much influenced by. (For critical reflections on analytic philosophy on the topic of delusions, see Sass 2004c.)

But even continental philosophy tends to be so abstract and sometimes overly historical in its focus. I prefer talking about something in the—something that's real, you know, like psychopathology for example.

MF: Or a person.

LS: Yes, a person. Or, groups or types of people or something like that. So I feel more legitimate because of that. I guess I'm enough of an anti-intellectual intellectual to feel that I would—I would have a lot of doubts about what I was doing if I'd gone into some of those other fields. I like the sense that this, clearly, is something real we are talking about, and ultimately, you know, even of practical importance—even if the latter has not been my own immediate focus.

MF: I sympathize with that response. You know, students will ask, why didn't I go into philosophy? And my simple answer is often that rather than studying "man" or "consciousness" or "Being," I'm interested in *this* person and to see what can be extracted from *this* life with all the tools that philosophy and psychology and anthropology and all the rest bring. We just have a minute or two.

About the discipline, it's still fraught. In many ways, I think you'd probably agree, it's stopped short of realizing its potential. What might the discipline of psychology do to somehow right its course? And I would ask that in two ways. What should it be doing that it's not, and what shouldn't it be doing that it is?

LS: Well, first of all I think maybe, at least in the field of psychopathology, which brings together clinical psychology, psychiatry, philosophy, cognitive science, and some other fields, I think there's reason to think that things *are* getting better.

MF: Good.

LS: The phenomenological movement of phenomenological psychopathology is actually having some—compared to the past anyway—having significant impact now, especially outside the U.S.A.. There are quite a few people now working on phenomenology, in philosophy, psychiatry, and psychology (among the most prominent I might mention Josef Parnas, Matthew Ratcliffe, Giovanni Stanghellini, Thomas Fuchs, and Iain McGilchrist); and some of us are also working on intersections between phenomenology and neurobiology, or thinking about it. So, maybe in some sense, some part of the profession of psychology, at least the part that has to do with mental illness, I won't say it's righting its course because this is still very much of a minority movement . . . but our minority movement is certainly gaining more of a voice.

MF: Yes, I noticed that from even the places where you're publishing your stuff. They're in schizophrenia journals, you know.

LS: Right, some of it.

MF: Truthfully, as opposed to theoretical and philosophical psychology journals.

LS: Some of it, yes.

MF: Or not to the exclusion of, but in addition to.

LS: Right, and other people as well are publishing in those kinds of journals. And so we phenomenologists, we phenomenological psychopathologists anyway, are being heard, to some extent—perhaps especially in Europe and Latin America; how much of an influence we're actually having on the Anglophone mainstream, I don't know. The mainstream in psychiatry, though, is pretty moribund. Some of you may know that the head of the NIMH, Thomas Insel, a few years ago basically said that there had been no real progress in research on or treatment of severe mental illness for the last several decades. And that's the head of the NIMH—former head now. Now, his own prescription for that is not phenomenology, it's more neurobiology.

But, at least there was a recognition, an acknowledgment that things are not progressing as one might have hoped or expected. So, the fact that one is not part of the mainstream is, under these circumstances, perhaps something to be a little bit proud of. But the question is: how does psychology right its course?

MF: Grow, flourish, become more of what it might be for our students . . .

LS: Well, I think being more open to so many of the things that, well, this conference, "Psychology and the Other," is all about, that's what psychology has got to do. I mean, my thing is phenomenology, but that's not the only thing, obviously. Forms of psychodynamic or psychoanalytic psychology, which have been losing their influence within the academy, need to be recuperated, recognized for their importance; there also needs to be more true openness to other fields like cultural anthropology and literary theory, and some parts of philosophy of course, and to a much richer interdisciplinarity overall. All that is very important.

I know that I'm preaching to the choir here, I'm sure everyone in this room will agree that the scientism of psychology is a considerable problem. And, you know, a lot of people—a lot of hard-nosed people recognize that. For just one such example, there's that great statement from Richard Feynman about "cargo cult science," do you know that one?

MF: No.

LS: This is what he thought of psychology. He called psychology "cargo-cult science" because, he said, it's like a cargo cult, in which people . . . (whether, from an ethnographic standpoint, this is really true of what cargo cults do is beside the point, by the way). Well, apparently in some South Sea Pacific islander cultures, they had gotten used to Western goods which had been brought to their islands by the navy and air force, and so, after World War II they wanted to get the planes to land on their islands, and to do so they would build fake runways, visible from above, and would wear headphones carved out of wood and bamboo antennae, and then hope that the planes would land so they could get all the stuff—the material goods—that they had come to appreciate.

Richard Feynman, the great physicist, describes all this, and then says that's kind of what psychology looks like to him. We psychologists, according to Feynman, we seem to follow all the trappings of experimental method, the ostensible experimental procedures of the hard sciences—at least overtly—but, as Feynman put it, for us at least: "The planes never land. The planes never land" (Gleick 1988). So, you could ask: am I anti-science or even anti-measurement?

MF: No, no, no.

LS: Of course not. But I do think, for example, that for neurobiological research to progress on something like schizophrenia or psychotic disorders more generally, there has to be a sophisticated awareness of what the correlates in subjectivity might be to the things being studied on the neurological plane. (On phenomenology's relevance for explanation as well as description, see Sass 2010, 2014b; Sass and Borda 2015.) Neurobiology may be doing lots of interesting things, but it is not making nearly as much progress as is often claimed, at least regarding mental illness—which was clear from what Thomas Insel said. And this was also clear to me when I went back to my neurobiological appendix in preparing the revised edition of *Madness and Modernism* (which will appear in 2017): I did of course have to update some things, but basically I found that not *all* that much has changed. The essential—the essential things being said 25 years ago about the neurobiology of schizophrenia, at least the *interesting* ones, are pretty much the same. The labels have changed a bit, some details have been added; but there are not a lot of new, revelatory ideas, that's for sure.

So, you know, contrary to the hype, there hasn't been all that much progress recently in neurobiology, at least regarding severe mental illness. But in my view, for there to be interesting progress, there needs to be a combination. There needs to be what the philosopher Shaun Gallagher nicely termed a relationship of "mutual enlightenment" between phenomenology and neurobiology. So, I think that that's something that we need to work toward.

By the way, I'm not trying to fetishize neurobiology or neurobiological research as some sort of ultimate standard, as if we had to measure ourselves by our relevance for neurobiology, not at all—I'm just giving it as an example, as a way of saying that even doing *that* will require a sophisticated appreciation of subjectivity.

MF: The distinction you've made between scientism and science is particularly important and it led me a short while ago to a kind of paradox. And that is that psychology in its scientism has paradoxically and ironically not been scientific enough

LS: Right. Amen. Amen.

MF: So any final comments, Louis, before we conclude? Did we miss —we probably missed some stuff.

LS: Well, I wrote a lot of notes based on some of your advance questions and we didn't get to all of those issues—but, of course, that's fine. It's impossible to get to everything. Still, there are a few points, some of them brought up by members of the audience here, that I would like to address, if I may.

One issue was brought up by Brent Slife, who is sitting here in the audience; and this is the question as to how we can understand others, even others who are not so obviously different from oneself as may be the case with schizophrenia. Brent suggested that, not only people with schizophrenia, but also normal individuals, even persons one might know very well, such as one's spouse, can also come to seem enigmatic; and he said that he had drawn inspiration from my interpretation of schizophrenia for the understanding also of people who do *not* have schizophrenia.

This made me reflect a bit of the whole question of different and same—which of course is always complicated. There are ways, certainly, in which schizophrenia can involve a radically different point of view from the perspective within which most of us live our lives. And Karl Jaspers was the person who insisted that it was, in fact, so extreme you couldn't even hope to understand it.

But there's another sense in which much that occurs in schizophrenia is also all very familiar—so long as one learns to recognize those aspects or those affinities. (One way of doing so, I argue, is to contemplate the phenomena of modernism and postmodernism—as a way of understanding the self-undermining and alienating effects that hyperreflexivity can have.) But, of course, one cannot say everything all at once, so there are times when one needs to emphasize the radical *difference* between the normal or commonsense position and that of schizophrenia, but also other times when one has to emphasize the other side, namely, the *similarities* or affinities. The latter, I think, is what Harry Stack Sullivan was getting at with his statement that people with schizophrenia are "more human than otherwise."

Brent Slife pursued his point by noting that, in fact, as another person comes to be better known, more familiar, one may actually come increasingly to recognize the possibility of a dramatic difference, perhaps even an unknowability; and he suggested that awareness of this difference, of this perhaps insurmountable gap, can actually be a crucial prerequisite for intimacy in general.

Brent was making (and, I am sure, was well aware of making) what could be described as a deeply Levinasian point: noting that you can find yourself falling into the abyss of an infinity that you cannot totalize (the philosopher Emanuel Levinas's most famous book is titled *Totality and Infinity*). And that, of course, is very true. To think you have *totally* understood someone else is probably to prove that you have missed something important. But I do think a lot of people would agree that there are *also* many tendencies in other people that are either very similar to our own reactions, or else that we may recognize from clinical knowledge. I think it's hard to deny, for instance, that very narcissistic people turn out to be very predictable in certain ways, all too predictable, in fact. Also borderline individuals, or with paranoid or obsessional traits or character styles, just to mention some additional examples. When I was much younger, I was very skeptical about all these categories. I remember, in fact, that when I first heard about borderline personality disorder and theories about its psychological structure, I almost laughed out loud; I just thought: how utterly absurd. How could you possibly think you could make these generalizations about people? It just seemed totally ludicrous to me.

It doesn't anymore, alas. I am afraid I would say that I have found that there just *are* quite a few people who fit into, who resemble these ideal types—remarkably so. And in that sense, we can sometimes know almost too much, and too readily about other people—especially as we get older and more experienced. Sometimes, in fact, it's almost scary how predictable other people can be. Which probably goes for myself as well, I realize. . . . But I don't think this contradicts what Brent Slife was saying, either. I mean, these are both truths that have their place.

Brent made still another point, asking whether I would agree that the people who are *not* considered schizophrenic might be described just as being people who are able to collude together in a kind of a rational or seemingly rational realm, and that people with schizophrenia just seem to have less ability to do that, or less tendency to do that. He suggested that what we define as sanity is really more a kind of collusion, a way of pretending that we are part of the same world when, in fact, we're not.

Brent's remark reminded me of a famous line from Jacques Lacan about psychosis, which is "*les non-dupes errent*"—a line that means, liberally translated, that those who are *not* duped wander lost and in error (Sass 2009). Roughly it means that. And so the people with schizophrenia are—they're not duped. In other words, often they don't buy into all of the conventional views that the rest of us buy into, and may need to accept in order to survive.

But as a result of *not* being duped, in the ways all the rest of us are, they end up being lost. Because it's not just a matter of not buying into a conventional belief or a prejudice here and there: language itself is called into doubt, and so are virtually all conventions and conventional concepts—and with that the very possibility of social cooperation and of many forms of valid thinking. And I think it's an incredible sentence, Lacan's, because it brings together the fact that there is indeed a kind of true insight, in a certain sense of that word, that people with schizophrenia can have, probably more acutely than the rest of us, but that, at least partly as a *result* of that, they're also wrong in some way—and lost in many ways. Well, it's just one sentence; also it's a pun in ways that make it even more complicated, but that's enough to make my point here.

Another person who is sitting in the audience, a man who identified himself as a psychoanalyst, asked whether I would agree that what we designate as psychotic is in fact highly context-dependent and perhaps culture-specific: if Jesus were now to be found walking about in the world, he said, would he not be viewed as psychotic and taken away by the EMT? My interlocutor suggested that there might be no essential structure, no "thingness" to schizophrenia that is invariant and transcends any social or cultural context; and he reminded us that various studies over the years have indeed demonstrated rather poor reliability in the application of the diagnosis.

This, of course, is also a highly complicated issue, and one that raises many questions. One thing to note is that, at least with the recent DSMs, you can in fact get rather good inter-rater reliability, meaning agreement between those making the diagnosis—so long as they truly stick to the criteria offered. But it seems that you may have terrible *validity*. So it may not be so much that professionals disagree when they use these criteria, it's that the individuals who are considered as having schizophrenia according to the current DSMs, the most recent DSMs, may not always be the patients that *should* be so designated. As is well known, some decades ago an embarrassing study demonstrated massive discrepancy in the application of the diagnosis of schizophrenia in the US and the UK, and this led to a tremendous focus on achieving interrater reliability, culminating I believe in DSM III. But it seems that such reliability was achieved at the expense of validity—with validity more or less thrown out the window in favor of clear-cut criteria that strove to minimize the role of clinical judgment. In a way, there may have been a more subtle and more valid diagnostic conception prior to DSM III, at least in Europe or parts of Europe.

But the real question is, I believe: is there any kind of essence to the condition or set of conditions that we tend to group together under the term schizophrenia? Does it make any sense to try to seek or attempt to define a kind of psychological essence to schizophrenia? And I guess my answer to that question is that yes, I think we do need to do that, or at least to attempt to do so—albeit not in a blind or rigid fashion. (As we explore it, we may need also to redefine the diagnostic boundaries—as often happens in scientific investigation.) And as you know, I've been one of several people suggesting that we might define schizophrenia in terms of a disorder of so-called "ipseity," which means a disorder of *minimal* or *basic* self that can be characterized fairly specifically by a combination of what we have termed "hyperreflexivity," "diminished self-presence," and "disturbed grip or hold" on the perceptual and cognitive world. (My work on self-disturbance in schizophrenia goes back a long way, and has developed in various ways over the years; see Sass 1987a; 1987b, 1992b, Chapter 7, 2000, 2001b, 2003b, 2014c, in press; Sass and Parnas 2001, 2003.)[2] Yet to say that there is an

[2] In Sass (2014c, p. 6), these concepts are defined as follows:

The self or ipseity disturbance in schizophrenia is hypothesized to have two main aspects that may seem mutually contradictory but are in fact interdependent. "Hyper-reflexivity" refers to an exaggerated self-consciousness, a tendency (fundamentally non-volitional) for focal attention to be directed toward processes and phenomena that would normally be "inhabited" or experienced (tacitly) as part of oneself. "Diminished self-affection" [a.k.a. as "diminished self-presence") refers to a decline in the (passively or automatically) experienced sense of existing as a subject of awareness or agent of action. . . . It is difficult to determine whether hyper-reflexivity and diminished self-affection are best conceived as complementary facets or tightly interacting processes; perhaps both conceptions are needed.

A third, interrelated aspect is a concomitant disturbance of the field of awareness labeled "disturbed hold" or "grip" on the world. Disturbances of spatiotemporal structuring of the world,

essence that's clear-cut and fully invariant over time has a ring to it that I probably wouldn't agree with. I often refer to the idea that—and I often cite Schwartz and Wiggins on this (the psychiatrist Michael Schwartz is sitting here with us in the audience)—that we should think of diagnosis in terms of something more like Weberian ideal types. (See Parnas et al. 2013; Wiggins and Schwartz 1991) And I think that, if you make that move, you can have a much more sophisticated and much more defensible way of talking about what schizophrenia is or might be. Does it have absolutely sharp boundaries? No, not any more than "capitalism" or "socialism" or "charisma" do. [These are classic examples of "ideal types" for the sociologist Max Weber, who formulated the notion of ideal types.] But is it completely a fantasy? No, just as capitalism, socialism, and charisma are not fantasies, but things that exist in the real world—things we need to recognize and to investigate, even if they can vary over time and perhaps sometimes even overlap with related notions in some respects

And so that's the kind of direction that I would go. And, as you may know, there are also attempts to operationalize the notion of ipseity disturbance through an interview technique, and so on and so forth (Nordgaard et al. 2013; Parnas et al. 2005). We have recently developed a similar semi-structured interview, known as the "EAWE: Examination of Anomalous World Experience," that is intended to focus less on self than on aspects of the lived-world, including the experience of space, time, language, and other persons (Sass et al. 2017). Also we have suggested ways of studying what might well be the neurobiological correlates of hyperreflexivity, of diminished self presence, of all of these phenomenological hypotheses. (On the neurobiology of self-disorder in schizophrenia, see Borda and Sass 2015; Nelson et al. 2014a, 2014b; Sass 1992b, appendix, 2014c, 2017, appendix; Sass and Borda 2015; Sass and Byrom 2015a.). We are also attempting to explore the notion by considering both affinities and discrepancies in the forms of abnormal self-experience that can be found in schizophrenia-spectrum as compared with other conditions, including mania and severe depression,

---

and of such crucial experiential distinctions as perceived-vs-remembered-vs-imagined, are grounded in abnormalities of the embodied, vital, experiencing self.

Depersonalization Disorder, the forms of dissociation found in Panic Disorder, and also intense forms of introspection. (See Madeira et al. 2017; Sass and Pienkos 2013a/2013b/2013c; Sass et al. 2013a, 2013b; also Sass and Pienkos 2013a/2013b/2013c.)

So, yes, I do think that ipseity-disturbance, this disturbance of minimal or core self-experience, should be considered something real: we should think of it as a real, albeit also problematic, multi-faceted, and ambiguous phenomenon that we're trying to study.

A final question that arises concerns the therapeutic implications of the phenomenological, or more specifically, of the ipseity-disturbance approach to schizophrenia. This too is a complicated issue, impossible to summarize here. I would refer any reader to several articles on that topic which I have helped to co-author (Nelson and Sass 2009; Pérez-Álvarez et al. 2010; Skodlar et al. 2013). One single point I would mention is the fact that one may need to beware of exacerbating hyperreflexivity through certain kinds of overly intellectual approaches to the patient. That can be a danger. But that's just part of it, and a lot of things that I and others in the phenomenological tradition would say about the treatment issue are quite congruent with what others writing about psychotherapy have said, for example, about the need to have a healthy relationship with the person with schizophrenia, to bring them back into the social encounter and so forth—but to do so in a way that is both non-coercive and genuinely respectful of both the nature and the potential validity of the patient's own, often idiosyncratic point of view.

MF: Let's thank Louis Sass for a fascinating set of ideas. Thank you so much.

Bibliography

Borda, J.-P., & Sass, L. (2015). Phenomenology and neurobiology of self disorder in schizophrenia: Primary factors. *Schizophrenia Research, 169*, 464–473.

Freeman, M. (1994). *Finding the muse: A sociopsychological inquiry into the conditions of artistic creativity*. Cambridge UK: Cambridge University Press.

Gleick, J. (1988, February 17). Richard Feynman dead at 69: Leading theoretical physicist. *New York Times*.

Heidegger, M. (1996). *Being and time*. Joan Stambaugh, Trans.. Albany, NY: SUNY Press. (Original work published 1927).

Laing, R. D.. 1960. *The divided self*. Harmondsworth, UK: Penguin.

Madeira, L., Carmenates, S., Costa, C., Linhares, L., Figueira, M. L., & Sass, L. (2017). Basic-self disturbances beyond schizophrenia: "Discrepancies" and "affinities" in panic disorder. *Psychopathology, 50*(2), 157–168.

McGilchist, I. (2009). *The master and his emissary: The divided brain and the making of the modern world*. New Haven, CT: Yale University Press.

Merleau-Ponty, M. (2012). *Phenomenology of perception* (D. A. Landes, Trans). London, UK: Routledge. (Original work published 1945).

Messer, S., Sass, L., & Woolfolk, R.Eds. (1988). *Hermeneutics and psychological theory: Interpretive approaches to personality, psychopathology and psychotherapy*. New Brunswick, NJ: Rutgers University Press.

Minkowski, E. (1927). *La schizophrénie*. Paris: Payot.

Nelson, B., & Sass, L. (2009). Medusa's stare: A case study of working with self-disturbance in the early phase of schizophrenia. *Clinical Case Studies, 8*, 489–504.

Nelson, B., Whitford, T. J., Lavoie, S., & Sass, L. (2014a). What are the neurocognitive correlates of basic self-disturbance in schizophrenia? Integrating phenomenology and neurocognition: Part 1 (Source monitoring deficits). *Schizophrenia Research, 152*(1), 12–19.

Nelson, B., Whitford, T. J., Lavoie, S., & Sass, L. (2014b). What are the neurocognitive correlates of basic self-disturbance in schizophrenia? Integrating phenomenology and neurocognition: Part 2 (Aberrant salience). *Schizophrenia Research, 152*(1), 20–27.

Nordgaard, J., Sass, L., & Parnas, J. (2013). The psychiatric interview: Validity, structure, and subjectivity. *European Archives of Psychiatry and Clinical Neuroscience, 263*, 353–364.

Parnas, J., Moeller, P., Kircher, T., Thalbitzer, J., Jansson, L., Handest, P., & Zahavi, D. (2005). EASE: Examination of Anomalous Self-Experience. *Psychopathology, 38*(5), 236–258.

Parnas, J., Sass, L., & Zahavi, D. (2013). Rediscovering psychopathology: The epistemology and phenomenology of the psychiatric object. *Schizophrenia Bulletin, 39*, 270–277.

Pérez-Álvarez, M., García-Montes, J. M., & Sass, L. (2010). *La hora de la fenomenología en la esquizofrenia* [Time for phenomenology in schizophrenia]. *Clnica y Salud, 21*, 221–233.

Sass, L. (1985). Time, space, and symbol: A study of narrative form and representational structure in madness and modernism. *Psychoanalysis and Contemporary Thought, 8*, 45–85.

Sass, L. (1987a). Introspection, schizophrenia, and the fragmentation of self. *Representations, 19*, 1–34.

Sass, L. (1987b). Schreber's panopticism: Psychosis and the modern soul. *Social Research, 54*, 101–147.

Sass, L. (1988). Humanism, hermeneutics, and the concept of the human subject. In S. Messer, L. Sass, & R. Woolfolk (Eds.), *Hermeneutics and psychological theory* (pp. 222–271). New Brunswick, NJ: Rutgers University Press.

Sass, L. (1992a). Heidegger, schizophrenia, and the ontological difference. *Philosophical Psychology, 5*, 109–132.

Sass, L. (1992b). *Madness and modernism: Insanity in the light of modern art, literature, and thought.* New York, NY: Basic Books.

Sass, L. (1994a). Civilized madness: Schizophrenia, self-consciousness, and the modern mind. *History of the Human Sciences, 7*, 83–120.

Sass, L. (1994b). *The paradoxes of delusion: Wittgenstein, Schreber, and the schizophrenic mind.* Ithaca, NY: Cornell University Press.

Sass, L. (1996). "The catastrophes of heaven": Modernism, primitivism, and the madness of Antonin Artaud. *Modernism/Modernity, 3*, 73–92.

Sass, L. (1998). Ambiguity is of the essence: The relevance of hermeneutics for psychoanalysis. In P. Marcus & A. Rosenberg (Eds.), *Psychoanalytic versions of the human condition and clinical practice* (pp. 257–305). New York, NY: New York University Press.

Sass, L. (2000). Schizophrenia, self-experience, and the so-called "negative symptoms." In D. Zahavi (Ed.), *Exploring the self: Philosophical and psychopathological perspectives on self-experience* (pp. 149–182). Amsterdam, The Netherlands: John Benjamins Publishing.

Sass, L. (2001a). Deep disquietudes": Reflections on Wittgenstein as antiphilosopher. In J. Klagge (Ed.), *Wittgenstein, biography and philosophy* (pp. 98–155). New York, NY: Cambridge University Press.

Sass, L. (2001b). Self and world in schizophrenia: Three classic approaches. *Philosophy, Psychiatry, Psychology*, *8*(4), 251–270.

Sass, L. (2001c). Wittgenstein, Freud, and the nature of psychoanalytic explanation. In R. Allen & M. Turvey (Eds.), *Wittgenstein, theory and the arts* (pp. 254–295). London, UK: Routledge.

Sass, L. (2003a). Incomprehensibility and understanding: On the interpretation of severe mental illness. *Philosophy, Psychiatry, Psychology*, *10*(2), 125–132.

Sass, L. (2003b). Self-disturbance in schizophrenia: Hyper-reflexivity and diminished self-affection. In T. Kircher & A. David (Eds.), *The self in neuroscience and psychiatry* (pp. 242–271). Cambridge, UK: Cambridge University Press.

Sass, L. (2004a). Affectivity in schizophrenia: A phenomenological perspective. *Journal of Consciousness Studies*, *11*, 127–147.

Sass, L. (2004b). "Negative symptoms," commonsense, and cultural disembedding in the modern age. In J. Jenkins & R. Barrett (Eds.), *Schizophrenia, culture, and subjectivity: The edge of experience* (pp. 303–328). Cambridge, UK: Cambridge University Press.

Sass, L. (2004c). Some reflections on the (analytic) philosophical approach to delusion. *Philosophy, Psychiatry & Psychology*, *11*(1), 71–80.

Sass, L. (2007). Contradictions of emotion in schizophrenia. *Cognition and Emotion*, *21*, 351–390.

Sass, L. (2009). Madness and the ineffable: Hegel, Kierkegaard, Lacan. *Philosophy, Psychiatry, and Psychology*, *16*(4), 319–324.

Sass, L. (2010). Phenomenology as description and as explanation: The case of schizophrenia. In S. Gallagher & D. Schmicking (Eds.), *Handbook of phenomenology and the cognitive sciences* (pp. 635–654). Berlin, Germany: Springer Verlag.

Sass, L. (2014a). Lacan, Foucault, and the "crisis of the subject": Revisionist reflections on phenomenology and post-structuralism. *Philosophy, Psychiatry, and Psychology*, *21*(4), 325–342.

Sass, L. (2014b). Explanation and description in phenomenological psychopathology. *Journal of Psychopathology*, *20*, 366–376.

Sass, L. (2014c). Self-disturbance and schizophrenia: Structure, specificity, pathogenesis (current issues, new directions). *Schizophrenia Research*, *152*, 5–11.

Sass, L. (2014d). Delusion and double bookkeeping. In T. Fuchs, T. Breyer, & C. Mundt (Eds.), *Karl Jaspers' philosophy and psychopathology* (pp. 125–147). New York, NY: Springer.

Sass, L. (2015). Lacan: The mind of the modernist. *Continental Philosophy Review*, *48*(4), 409–443. doi:10.1007/s11007–015–9348-y.

Sass, L. (2017). *Madness and modernism: Insanity in the light of modern art, literature, and thought: Revised edition*. Oxford UK: Oxford University Press.
Sass, L. (in press). Schizophrenia as a self-disorder: Theoretical clarifications, recent developments, new directions. In G. Stanghellini, M. Broome, A. V. Fernandez, P. Fusar-Poli, A. Raballo, & R. Rosfort (Eds.), *The Oxford handbook of phenomenological psychopathology*. Oxford, UK: Oxford University Press.
Sass, L., & Borda, J.-P. (2015). Phenomenology and neurobiology of self disorder in schizophrenia: Secondary factors. *Schizophrenia Research, 169*, 474–482. doi:10.1016/j.schres.2015.09.025.
Sass, L., & Byrom, G. N. (2015a). Phenomenological and neurocognitive perspectives on delusion. *World Psychiatry, 14*(2), 164–173.
Sass, L., & Byrom, G. N. (2015b). Self-disturbance and the bizarre: On incomprehensibility in schizophrenic delusions. *Psychopathology, 48*, 293–300.
Sass, L., & Parnas, J. (2001). Phenomenology of self disturbances in schizophrenia: Some recent research directions and findings. *Philosophy, Psychiatry, Psychology, 8*(4), 347–356.
Sass, L., & Parnas, J. (2003). Schizophrenia, consciousness, and the self. *Schizophrenia Bulletin, 29*, 427–444.
Sass, L., & Pienkos, E. (2013a). Delusions: The phenomenological approach. In W. Fulford, M. Davies, G. Graham, J. Sadler, & G. Stanghellini (Eds.), *Oxford handbook of philosophy of psychiatry* (pp. 632–657). Oxford, UK: Oxford University Press.
Sass, L., & Pienkos, E. (2013b). Space, time, and atmosphere: A comparative phenomenology of melancholia, mania, and schizophrenia, Part II. *Journal of Consciousness Studies, 20*, 131–152.
Sass, L., & Pienkos, E. (2013c). Varieties of self experience: A comparative phenomenology of melancholia, mania, and schizophrenia, Part I. *Journal of Consciousness Studies, 20*, 103–130.
Sass, L., Pienkos, E., & Nelson, B. (2013a). Introspection and schizophrenia: A comparative investigation of anomalous self experiences. *Consciousness and Cognition, 22*, 853–867.
Sass, L., Pienkos, E., Nelson, B., & Medford, N. (2013b). Anomalous self-experience in depersonalization and schizophrenia: A comparative investigation. *Consciousness and Cognition, 22*, 430–431.
Sass, L., Pienkos, E., Skodlar, B., Stanghellini, G., Fuchs, T., Parnas, J. & Jones, N. (2017). Examination of Anomalous World Experience (EAWE). *Psychopathology, 5*(1), 10–54.

Sass, L., & Woolfolk, R. (1988). Psychoanalysis and the hermeneutic turn: A critique of narrative truth and historical truth. *Journal of the American Psychoanalytic Association, 36*, 429–454.

Shapiro, D. 1965. *Neurotic styles*. New York, NY: Basic Books.

Skodlar, B., Henriksen, M. G., Sass, L. A., Nelson, B., & Parnas, J. (2013). Cognitive-behavioral therapy for schizophrenia: A critical evaluation of its theoretical framework from a clinical-phenomenological perspective. *Psychopathology, 46*, 249–265.

Wiggins, O. P., & Schwartz, M. A. (1991). Research into personality disorders: The alternatives of dimensions and ideal types. *Journal of Personality Disorders, 5*, 69–81.

Mark Freeman is Professor and Chair of the Department of Psychology and Distinguished Professor of Ethics and Society at the College of the Holy Cross in Worcester, Massachusetts. His writings include *Rewriting the Self: History, Memory, Narrative; Finding the Muse: A Sociopsychological Inquiry into the Conditions of Artistic Creativity; Hindsight: The Promise and Peril of Looking Backward; The Priority of the Other: Thinking and Living Beyond the Self;* and numerous articles on issues ranging from the psychology of memory and identity to the psychology of art and religion. Winner of the 2010 Theodore R. Sarbin Award in the Society for Theoretical and Philosophical Psychology, he is also a Fellow in the American Psychological Association and serves as editor for the Oxford University Press series "Explorations in Narrative Psychology."

History, Morality, and the Politics of Relationality: A Conversation with Philip Cushman
(Interviewed by Heather Macdonald, Lesley University)

In 2003 Dr. Philip Cushman, well into his career, taught a course for the graduate program in existential phenomenological psychology at Seattle University. Dr. Steen Halling, a beloved professor of the program, was on sabbatical and Dr. Cushman had been hired to cover his class for the semester. At this time I was a first year student of the program and disgruntled to find that there would be a new faculty member teach Dr. Halling's course. Now, of course, I realize how lucky I was to have had such an experience. I have a particular memory from the class as I began to realize the magnitude of the ideas we explored. As I was sitting in class, I looked briefly through the window and could see springtime burst forth outside; the emerald leaves came forth from their buds on the branches and the grass shimmered in the sunlight. We talked about culture and the history of psychotherapy and these textured geographies of moral selfhood.

P. Cushman (✉)
Antioch University Seattle, Seattle, WA, USA
e-mail: pcushman@centurytel.net

H. Macdonald et al. (eds.), *Dialogues at the Edge of American Psychological Discourse*, Palgrave Studies in the Theory and History of Psychology, DOI 10.1057/978-1-137-59096-1_4

We explored the linkages between history, hermeneutic traditions and underground flows of power that shape subjectivity—and I felt like I was home. I had a sense of being home for the first time.

So I am very grateful for taking the class and for Dr. Cushman's scholarship. As someone who considers herself more of an anthropologist than a psychologist, your work really spoke to me in a very deep way. Eric Severson said earlier in the conference that a single course with Cushman is like an entire degree elsewhere. I couldn't agree more, and that's exactly what my semester at Seattle University was like.

As many of you know, Dr. Philip Cushman is an esteemed scholar, teacher, psychotherapist whose contributions to the field are difficult to catalogue in this short introduction. Dr. Cushman holds a Master of Arts degree in American Studies, and a Master of Arts degree in Family Therapy, and the Ph.D. in Psychology. He served as core faculty in the doctoral psychology department at the California School of Professional Psychology in Alameda, California and core faculty in the doctoral psychology program at Antioch University in Seattle.

In 1995, Dr. Cushman published his first book, *Constructing the Self, Constructing America: A Cultural History of Psychotherapy* in which he provides a history of psychotherapy from critical, Foucauldian and hermeneutic perspectives by building on the work of T. J. Jackson Lears, Christopher Lasch, Edward Sampson, Martin Heidegger, Hans-Georg Gadamer, and Kenneth Gergen. In addition to this work, he has published over 50 articles on a wide range of topics that are drawn from the disciplines of history, philosophy and psychology. He has an ability to carefully synthesize these domains when crafting his own scholarship.

He has especially relied on hermeneutics to develop a critical perspective on the cultural role of psychotherapy in order to improve its practice. He argues for an interpretive approach to social science that recognizes no single or simple truth, but truths that must be arrived at through deep dialogue and with what he calls an encounter with difference that is fully embraced.

He believes that humans are always in relation to one another, to society, and to the historical traditions that constitute us, and that psychotherapy is inevitably a moral practice in a political context. Dr. Cushman's most recent publication includes a 2015(c) article titled,

"Relational Psychoanalysis as Political Resistance," in *Contemporary Psychoanalysis* and he is at work on a second book, hopefully we'll be able to talk a little bit about that. And lastly, this past summer he received the APA Division 24 Award for Distinguished Lifetime Contributions to Theoretical and Philosophical Psychology.

Heather Macdonald (HM): To begin with, in one of our earlier conversations this summer, when I asked you how the field of psychology responded to your work, you looked at me and shook your head and said, "You've got the wrong guy." So, why don't we start there and perhaps you can explain your initial reaction.

Philip Cushman (PC): First I want to say how touched and honored I am by this invitation. Flabbergasted by it, really. I was extremely uncomfortable about it for months after I heard about it. Now I can say I'm not extremely uncomfortable any longer, I'm just uncomfortable. [laughter] And I'm really looking forward to this being over, so I'm glad it's started, because then it'll be finished.

This invitation has been very challenging. I've been trying to figure out why I'm so uncomfortable about it, with somehow the hope that a little awareness could lower my discomfort. It looks like Dr. Daniel Masler, a former student of mine, is not present right yet, but I want to mention this to you: it must have been Thursday night, we were in a tavern near Harvard Square and Daniel asked me about this interview and I told him about how uncomfortable I was. Right there in the bar Daniel took it upon himself to cure me of my nervousness. But nothing much worked.

I mention what came next because those of us in the room who are therapists know that calculated interventions almost never work—much of my writing these days focuses on the wrong-headedness of the manualization of psychotherapy and the structuring of doctoral psychology education through academic competencies. It's what you say or do, when you're not planning it or maybe even thinking about it, that seems to help somehow—Donnel Stern, in a brilliant article in 1990, called the secret to successful psychotherapy "Courting Surprise." So, we're sitting in this noisy bar and Daniel delivered this offhanded remark: "You know, I don't understand what you're concerned about. This interview will be just like teaching, and you love teaching." And it was that one spontaneous remark that helped a lot. I was kidding about "cure," as I hope most of you know, because cure is an especially problematic concept. In fact, in 2011 a paper of mine titled "The Danger of Cure, the Value of Healing" was published as a chapter in a 2012(a) book titled *Midrash and Medicine*. The way Daniel and I used "cure" was just as a joke.

Still, truth be told, I'm not comfortable with sitting before you today because I don't think I've done anything that warrants this kind of attention. I didn't set out to make a big contribution, I just set out to help my family survive, which maybe I can tell you more about that as time goes by here today.

Part of my discomfort also may be because I'm angry about our profession, especially the more acontextual, ahistorical, aphilosophical trends we have witnessed over the last 30 years or so, and the arrogance with which the profession now presents itself to the public, especially the claim that psychology is a STEM discipline. Maybe I don't want to be seen because I'm sure other psychologists won't agree, and I don't want sharp objects thrown at me. Ah, here's my therapist now! [laughter as Dr. Masler enters the room] Daniel, I told them that you cured me in one session at the tavern the other night.

Dr. Masler: Yeah, but I can't disclose the specific technique I used. [laughter]

PC: Fortunately, however, a book is forthcoming. What was I saying? I think my anger and disappointment and grief and fear make it possible for me to write. Along with teaching, writing is certainly the most gratifying thing I do.

HM: I'm curious about what drew you to the field of psychology to begin with, rather than American Studies. And what or who influenced your early entry into the discipline. I'm thinking of people like Charles Taylor and Ed Sampson. I'm also thinking of the influence of Harry Stack Sullivan and Heinz Kohut.

PC: Okay. But allow me a moment before we delve into that. I want to say that I have a great deal of gratitude for the Psychology and the Other Institute and the director, David Goodman, and you, Heather, and all the folks who work for the Institute. Looking back on my life, I feel like a very lucky person. Somehow I've mostly been able to teach courses I want to teach and in the ways that I want to teach them. And I've been blessed with wonderful students, who I love. It's been the most remarkable experience of my life to be able to teach—my first full-time teaching job was in 1972—and to be involved with all of them.

I feel a great deal of gratitude toward my friends in psychology and especially friends in Division 24 and most especially to Frank Richardson and Blaine Fowers and more recently, Robert Bishop, for teaching me and helping me avoid the "postmodern blues," (that's what we called it back then) by teaching me more about hermeneutics, which over time has meant a great deal to me and my work (Cushman, 1995b; See references for additional works). I couldn't have begun to write the way I do without my colleagues working in the field of hermeneutics. It's through their challenging questions that I really did learn about Taylor and Gadamer, whose work helped me understand Heidegger better.

What you didn't mention about the beginning of my academic life were two graduate programs I was in before I went into American Studies. I didn't finish either one, although they had a profound effect on the way I think and live. The first was a graduate rabbinic program, beginning in 1967, which I left in early 1971. In addition I was in a

UCLA Folklore and Mythology program for a couple of semesters before I finally figured out that what I loved about all of these programs was history. And that's when I left and went into the American Studies program with a concentration in history.

I realize now, after reflecting on your questions, that leaving the rabbinic program threw my life into chaos in a way that at the time I couldn't really let myself notice. It had been about the only economically safe place I could imagine myself in.

Leaving there threw me into a world that I didn't really understand. Karen and I got married at about that time. Both of us come from working/lower middle class families, and we didn't really know anything about the university; we were so ignorant about the politics of university culture, and I just kind of stumbled along. Although I knew I loved teaching—that was the one thing I knew. So I just stumbled from place to place. Many of the therapists I know seemed to know at a young age where they belonged, what they wanted to do, and how to get there. But that was never the case for me. I was just hoping somehow to scrape up enough money for us, and when history jobs became so difficult to get, I turned to psychology out of necessity. My wife wasn't working much at that time and we had a baby daughter named Leah Corinna Katie Shoshanna Starshine Cushman—guess in which era she was born. [laughter] And I was just desperate. Although I wouldn't have admitted it at that time, I just kept working, hoping to find my way. I felt like, well, I'm interested in this subject. Maybe I can do it, but the reality of getting work in the humanities was grim.

After the first college I taught at collapsed for economic reasons, I called a former professor of mine who taught at Boston University and was the chair of what was then called Afro-American Studies. A couple of years before that he told me that if I ever wanted to go to B.U. in history I should call him and he could help me apply and get a fellowship. But in 1973 when I called, he said "Don't come. I can't even get jobs for my own graduates. History is dying. Don't do it." I left that phone call and felt devastated. After hustling adjunct teaching jobs for a year I decided to go into a marriage and family therapy program, a new field at that time and very hip. I hoped that I could get a quick M.A. degree and earn enough money for us. We were just hoping that agency work or a private practice would allow our young family to survive—it never occurred to me to write academic articles.

HM: Nevertheless, history as a discipline never died for you, because throughout all of your scholarship you relentlessly apply the work of Martin Heidegger and Hans-Georg Gadamer as a way to highlight the political and historical contexts of psychotherapy, especially as they pertain to clinical practice.

PC: So true; history had become an integral part of my life. I was always hoping for some kind of Community College teaching job that would allow me to combine psychology and history. But I could see the handwriting on the wall. It was no longer possible to get a college teaching job with only a M.A. degree. So I was thinking I'd just do clinical work for the rest of my life.

We moved to San Francisco, and after a couple of years it became clear to me that I was going to have to get the Ph.D. in order to keep a solid, full-time practice going. I decided to get a quick Ph.D. and I enrolled in what was then called Humanistic Psychology Institute [HPI]. I couldn't go to a better-known school because for financial reasons I couldn't give up my private practice.

I told Karen: "I promise, no more politics. I'm not going to even think about activism. I'm just going to get this degree and increase my practice and make money." So the first day I was at HPI, what these days is called Saybrook Institute, my first instructor was Ed Sampson, whose work I didn't know at all. During the break of that first class I called Karen and said, "Kar, I'm having a great time." She sounded a little suspicious, but I went on. "This is really interesting, I'm starting to learn about the political meanings of therapy." And she said, "Uh huh, and how long did it take?" I said, "About ten minutes." [laughter] Ed and I worked together in several classes and have become good friends. He was enormously helpful to me, helping me see connections between history and politics and psychology.

So that's how it started. Also around that same time I was working at Jewish Family Service, and I designed a program that helped parents rescue their grown offspring from religious and psychological restrictive groups. As part of my course work I learned about Kohut and later I wrote a couple of articles about restrictive groups, drawing from social psychology and a Kohutian perspective.

HM: Yes, Kohut's (1971) Self Psychology figures prominently in your earlier articles. As I was reviewing your very early work, I came upon an article that you wrote in 1987 titled, "History, Psychology and the Abyss: A Constructionist-Kohutian Proposal."

It seems that one of Kohut's ideas was that we could never really get to an essential self—or selfhood—but we could know these cohesive or constructive forms of self. So, was Kohut the beginning of your understanding of the self and then what led you to your later concepts of the empty self and the flattened self?

PC: I started thinking about constructionism from Ed Sampson and Ken Gergen (See this volume). And then I intensively studied Kohut in some clinical classes I took. Many of my therapy colleagues in the East Bay were enamored with Kohut. For instance the peer consultation group I was in was very Kohutian at that time, Kohutian and Interpersonal. So, out of necessity, I struggled to understand those theories because I didn't learn anything like that in family therapy. In the family therapy program, I mostly learned how to trick families into doing things that I wanted them to do. That's not an exaggeration, really, because systems and power were extremely popular right then—Jay Haley, Cloe Madanes and Ordeal Therapy being all the rage.

I started learning about Kohut before I started at HPI. My wife was in a Museum Studies program and got an internship at Sturbridge Village in Massachusetts, a wonderful, Open-Air museum. So I quit my job at Jewish Family Service and the three of us moved to Sturbridge for the summer. I wanted to study with Howard Zinn, but he was gone for the summer. However he was gracious to me over the phone and suggested I call John Demos, who was teaching then at Brandeis (now at Yale) and is a wonderful American historian who was one of the historians who studied under Kohut. I studied with John that summer, especially Kohut's ideas about narcissism, charismatic leadership, and groups.

It was John who introduced me to Chuck Strozier, another historian who had studied under Kohut, who was then the editor of *Psychohistory Review*. What was especially interesting to me about Kohut was his interest in history and how he talked about what he called the group self. Now, it turns out that's a pretty problematic concept. But the overall question about how a group can at times help a person psychologically hold together was interesting to me, especially given the work I was doing with people in restrictive groups.

That summer with Chuck's help I wrote a paper just for my own purposes about recruitment and indoctrination processes in religious and psychotherapy restrictive groups titled "The Politics of Vulnerability: Youth in Religious Cults." I was just about to start the doctoral program then and after a couple of years and several courses in social and clinical psychology and some rewrites, much to my surprise, *Psychohistory Review* (1984) published the damn thing! I had found a way of combining some of the social psychology I was learning at school with Kohut and his ideas about the self, and related all that to what I already knew about cults. I hadn't set out to write an academic article, but I got interested in the subject. In response to your initial question, Heather, the part of the article I liked writing and thinking about the most was the last little section that was a brief historical analysis of the post-WW II cohort—my first initiation into psychohistory. In 1986, with help from my dissertation chair Susan Hales, I published a similar article titled "The Self Besieged (1986)" in *Journal for the Theory of Social Behaviour*. It was a contribution to a special issue on "The Rediscovery of Self in Social Psychology." All of that eventually turned into a dissertation titled "The Politics of Transformation: Recruitment and Indoctrination Processes in a Mass Marathon Psychology Organization (1989)." In the process of that research I read Louis Sass (See this volume) for the first time; he was using Heidegger and drawing from history to critique psychotherapy theories, and I thought his work was brilliant. With Louie's work as a guide, I began trying to combine constructionism, history, and Self Psychology; out of that ungodly mix came my article "History, Psychology, and the Abyss" in 1987. You could tell I didn't really understand the full implications of

Heidegger at that time, and to tell you the truth I was trying desperately to save Kohut from the implications of the more rigorous hermeneutic critique I was beginning to learn more about.

As I look back on it now I realize I was trying to understand some of the problems with Self Psychology, especially the way some of Kohut's followers had a tendency to treat his theory in such an ahistorical way, as if Kohut had found the one truth about the human psyche. "First," they thought, "Freud was good for a while, but the 'flaws' in his theory became obvious, then Sullivan's Interpersonal theory was good, but then also somewhat limited. However, now truly Kohut has delivered to us the one truth."

I was trying not to pay attention to the Kohut worship when I wrote the 1987 article. But it didn't really work because I could see that constructionism—and a fuller hermeneutic vision—was really calling that unlimited embrace of Kohut into question in a serious way. Fortunately, that article was the last time I tried to do that. [laughter]. And I continued learning more about Heidegger, and later, mercifully, Gadamer.

One month in 1988, my copy of *American Psychologist* arrived in the mail, and I read Sampson's article, "The Debate on Individualism," which I found very exciting. Of course, I had read many other articles by him, including especially the 1981 article "Cognitive Psychology As Ideology," which I just loved. I thought it was brilliant, and I still do, I teach it every year in the History and Systems of Psychology course and in fact in several other courses. I think it is even more applicable today than back then.

So, in 1988 I read "The Debate on Individualism," and I remember thinking to myself, "I think maybe I could write an article like this." I made some notes in between patient hours that day, and those notes slowly became the outline for "Why the Self is Empty."

HM: Take us through the development of that article because I think that's a foundation for your book in 1995 (a/c), *Constructing the Self, Constructing America: A Cultural History of Psychotherapy*.

PC: Yes, that's right. I had taken a couple of courses from Bert Dreyfus, who was a most important Heidegger scholar. I was first taught about Heidegger by a beloved teacher of mine and

> good friend named Tony Stigliano, who helped me think more about the self from a hermeneutic and historical point of view, and I started to get ideas. As I tried to understand Heidegger's concept of the clearing, I started to think about why Americans are so psychological now, in the decades after World War II: what does America as a psychological society mean about us and our time? And why are therapists obsessed with concepts about the self? I noticed that there was a magazine called *The Self* that had started publication. I realized that it wasn't only therapists who were so infatuated by the concept of the self, it was most everyone.

Soon after, I took a class from Donald Lowe, a Marxist historian at San Francisco State, an evening class in which my friend Terry O'Hare and I enrolled. Like me, Terry had graduated in history but couldn't continue. And so we were excited about this class, which was titled something about emotions, subjectivity, and the body.

At one point after class I went up to Dr. Lowe, who had written a wonderful book in 1983, *The History of Bourgeois Perception*, which broadly speaking was about the Victorian Era. He argued that in every historical era, one or two physical senses predominate and they fit with the basic communications medium of that era.

I started to wonder: what is the fit between our current era's predominant sense and basic communication medium? I started noticing that narcissism and borderline states—what are called Disorders of the Self—were becoming the diagnoses of our time. I noticed that often in case studies and informal conversations among therapists, emptiness had become a predominant symptom of most of our patients. I started noticing that there were commercials and ads that exploited a sense of emptiness and even encouraged it.

So I went up to Dr. Lowe once at the end of a class session and said to him, "From what I'm learning in this class and some reading I've been doing in Heidegger, I've got an idea that I think fits with what you are saying in the book. I think we're developing an empty self that fits with our economy, advertising, and pop culture, and I think it is reflected in psychotherapy theory." And he said, "That makes absolutely no sense to me, it's a ridiculous idea." And he walked away.

But I didn't walk away from that idea because it really was interesting to me and I wanted to write about it. So from Ed's *American Psychologist* article, I took courage and started the process of writing "Why the Self Is Empty," (published in 1990) and sent it in to the *American Psychologist.* Stan Messer, who was the action editor for the article, put together a wonderful review committee that, I found out later, included Charles Taylor, Robert Bellah, Jackson Lears, and Louie Sass—what a committee! Louie, God bless him, gave Stan permission to disclose Louie's identity to me, and I traveled to the American Psychological Association annual conference for the first time in 1989 because I wanted to meet Louie and talk to him about the manuscript, which they had returned with suggestions for a rewrite. He was generous with his time, patient with me, and very helpful.

I want to tell you a brief story about Charles Taylor. These reviews were all pretty formal and quite extensive, except for one that was, honest to God, just a few lines long. It was obviously written on a manual typewriter because the capitals weren't done properly. More or less he wrote, "Look, I'm not a psychologist. In fact I don't understand why I'm on this review panel, and I'm not sure of your criteria or exactly what you want. This paper is okay, I guess. It uses a somewhat limited, constructionist approach, which is not very nuanced. But I suppose if I have to say yes or no about it, I guess it'd be okay with me if you publish it."

I'm not implying in any way that his review was mean-spirited. I don't think he is that kind of person, and I certainly didn't take it that way. Now I realize that he was right about constructionism. I felt it then, but I couldn't exactly say why. So I went and met with Louie and he helped me create a more nuanced argument. Through Ken Gergen's (See this volume) writing I had stumbled upon a couple of other folks, like Jill Morawski, who were very helpful to me. I also met with Jill at APA in 1989 and we had a great time walking in New Orleans and talking for hours about history and psychology, which also helped me with the Empty Self manuscript. In other words, a lot of brilliant and learned people helped me enormously. They made it possible for the article to be good enough to be published.

So that's how "Why the Self Is Empty: Towards a Historically Situated Psychology" came into being.

HM: I'm assuming "Why the Self Is Empty" really laid the foundation for your 1995 book *Constructing the Self, Constructing America*. Can you speak briefly about how that article and others set the foundation, just so we can understand the link?

PC: In 1992 (a/b), I was asked to contribute a chapter to a book that APA called *History of Psychotherapy*. My chapter was titled "Psychotherapy to 1992: A Historically Situated Approach." After that I took parts of that chapter and the empty self article, the historical parts, condensed them some, added more technical sections about therapy, and expanded all that into the book.

In many ways, the book wrote itself, in part because it was historical and so mostly I just went era by era. I was then able to include a lot of wonderful ideas that I'd learned over the years from my hermeneutic colleagues, and also I was able to throw in some Judaica, which was fun for me and personally meaningful. I got through eight chapters and I knew what I was going to write for what finally became the tenth chapter, titled "The Politics of the Self," and my editor at Addison-Wesley said, "We want you to write another chapter. We want you to write about how you use hermeneutics in therapy."

And I said, "Oh, thanks for the suggestion, that's a great idea, but I couldn't possibly do it. I don't know how to do it. Nobody knows how to do it." The European psychiatrist Medard Boss, a contemporary of Heidegger's, tried to do it in 1963, but it didn't translate well to post-World War II America.

She said, "No, no, no, you don't understand," in the way that publishers say these things. "You don't have a choice. You don't write that chapter, we don't publish the book." They only did that twice during the years I wrote the book. The other time they overruled me was about the subtitle. I wanted the subtitle to read "Studies in a Cultural History of Psychotherapy." They said, "That sounds way too academic, we can't use that. Let's just call it "A Cultural History of Psychotherapy." And I had to do it their way. I'd say the best criticism of the book is that it's not a comprehensive cultural history. It was a compilation of *studies*. I've always wanted in public to say: "see, it's not my fault"—I thought since we're talking about this, I could throw that in!

HM: Yeah, you can throw that in. [laughter] So this really broke some ground in writing about how to use hermeneutics in therapy.

PC: Well, I don't know about that, but reluctantly I did set about trying to write that chapter. In the course of that work, which was pretty challenging—very challenging—I came to realize that in certain ways I was already doing therapy hermeneutically. That was comforting, and encouraging, although I'm sure I was just stumbling along, as we all did in one way or another, and still do, trying to apply a different, hermeneutic way of listening to, and thinking with, another person. In those days a few of us in this room were trying to use hermeneutics, but it was not easy, especially when other therapists in our geographical areas couldn't begin to understand what we were trying to do. So it was lonely work; that's why, for me anyway, the annual APA conferences were so important: I didn't feel so alone. Frank Richardson and Blaine Fowers and I would talk late into the night about hermeneutics and therapy, and those talks helped me have more confidence in what I was trying to do. Through my friends I found the courage—and the intellectual resources—to go on.

In the years after *Constructing the Self, Constructing America (1995)* I continued to think historically about therapy (See Cushman, 2002, 2004). Life would be so much easier if I didn't, but I can't seem to help myself, thinking historically is who I am. I began to notice what I thought might be a new type of self emerge, a new way of being. Peter Gilford (Gilford and Cushman, 2001), a student at the time, and I began to examine how managed care and computers were influencing Americans, and we wrote a paper and submitted it to *American Psychologist.* I thought it was an important piece, a companion to my 1990 article, but it was rejected. And they thought it was so weird they told us not to resubmit it. So Peter and I divided it in half, and turned the more historical part into an article for a 1999 special issue of *Psychohistory Review* called "The Self at the Year 2000." Our article was titled "From Emptiness to Multiplicity: The Self at the Year 2000." Then we reassembled, added to, and then resubmitted to *American Psychologist* the second half of the original article, now titled "Will Managed Care Change Our Way of Being?" With the help of Blaine Fowers, who gave us editorial suggestions, and Ken Gergen, as tireless action editor, the new article was (finally) accepted and published in 2000(b).

I was shocked that the first version was turned down, because it was an extremely important issue, and I thought we wrote about it in a creative and compelling way. But in retrospect, I now realize that, given the changing neoliberal cultural landscape of the country, and the direction in which APA was leading the field, our article would have represented a serious challenge to psychology's collusion with managed care. Without Ken's farsightedness and dedication to the project I don't think it would ever have been published. By the way, when our article appeared in *A.P.* it was not commented on by psychologists—I got very few comments about it from anyone. My hunch is that it was describing our dilemma as therapists in ways that were just too disturbing for most of us to think about, because then we'd have to face the way health care corporations and market values were affecting our work and our personal lives. In any event, more recently, we've received a bit more of a response, but still, nothing like the volume of letters and depth of feeling I received from the publication of the empty self article.

HM: So, as I see it, this leads us to your next phase of work. In 2005, you published an article called "Between Arrogance and a Dead End: Psychoanalysis and the Heidegger/Foucault Dilemma." I really love this work because it gives us a good summation of the key critical questions that you're working through. One of the major themes in your scholarship has been to address the constitution of local moral selves through Gadamer and within the context of larger historical forces as contextualized by both Heidegger and Foucault. However, more recently, you started to write about the limits of Heidegger's thought, the need for but limits of Michel Foucault's political vision, and the way Gadamer's concept of dialogue brings an essential element to their work and the dilemma it produces. I think that dilemma exists as a tension throughout your recent work. So could you talk about the tensions between Heidegger and Foucault—and then Gadamer as a solution between the tensions? In other words, what does Gadamer offer that Heidegger and Foucault cannot?

PC: And we are supposed to be finished in . . . how long? [laughter]

Well, the "Arrogance and a Dead-End" article came out of research that I did especially in the academic year 2000–2001, when mercifully I had my one and only sabbatical. During that year, I took another class from Bert Dreyfus on Heidegger and this was at University of California, so I could see how he conducts the Heidegger class each year, which was fascinating. I also took a class on Foucault, and those two courses were precious to me. Except for many (many) informal conversations, those were the only formal classes I ever had in hermeneutics. The rest of what I've learned I just cobbled together through conversations with friends and my own studies.

One thing that I got increasingly interested in as my sabbatical went on was Heidegger's involvement with the Nazis. Because I am a positively identified Jew I was always interested in the Shoah, and so Heidegger's actions, including his year-long membership in the Nazi party—and especially his refusal to renounce the Nazis after the war—bothered me enormously.

I remember once I went to a big international conference at Berkeley during the earlier stages of Heidegger's popularity in the U.S. and it became obvious that many Heideggerians were covering up Heidegger's Nazi involvement. It was pretty creepy. I remember soon after that, I talked to Charlie Guignon about it and I began to realize that there was much more there that I had to learn about and write about (See Cushman, 2015a).

So, I read many books on Heidegger's Nazi involvement, such as Richard Wolin's (1993), who argued that Heidegger's pro-Nazi actions implicated his entire philosophy, Hans Sluga (1993), who located Heidegger's anti-Semitism within the history of German philosophy, and Saul Freidlander (1993), who attributed it in part to 1900 years of Christian theology. Charlie helped me with this issue by elaborating on Gadamer (1960/1989), especially his concept of "dialogue." Charlie explained that just because hermeneuticists believe we are constituted by the historical traditions of our time, that doesn't mean that it falls into a kind of historical determinism. What we are doing continually in our lives is what multiculturalists sometimes call "encountering difference," and then allowing the moral understandings of the other to place our own understandings into question. In this approach, we are continually tacking back and forth between our beliefs, commitments, perspectives, and moral understandings and those of others. It is that tacking back and forth that helps us determine whose understandings of the good are best for that particular moment. That process is what helps us shift our perspective so that we're not just imprisoned by the status quo of our current social world. Gadamer called that process dialogue, or "genuine conversation" and I saw it as a way out of the Heidegger-Foucault dilemma. I still do. I think I'd be lost, or at least still very confused, without Taylor's explanation of the inescapable moral nature of historical traditions and Gadamer's concept of dialogue.

That's what I tried to write about in "Between Arrogance and the Dead-End," finally published in 2005(a). In a moment of chutzpah, I accused Heidegger of monoculturalism, which I still think is true. I came to understand that Heidegger had been influenced by Christian anti-Semitism and the previous 150 years of German philosophy, some of which argued that the Germanic people had a special destiny and in the near future would culturally dominate Europe and the world. As a result, he was unable to appreciate the contributions that non-Aryan people, such as the Jews, made to German culture. This in turn made it easier for him to overlook the Nazi's brutal anti-Semitic policies and in general the Nazi's crude and violent approach to political life. Heidegger's monoculturalism undoubtedly contributed to his delusions about the Nazis and especially his imagined place as *the* philosopher to Hitler. His withdrawal and isolation after the war probably produced new aspects of his philosophy that moved him away from the centrality of culture and relationality. All that is why I used arrogance in the title of the article.

At the same time I was struggling with Foucault's brilliant historical contributions. As I read Foucault's hermeneutic critics, I came to see that Foucault lacked an interest in or ability to talk directly about moral understandings and to realize how they are primary to the exercise of power. This is the hermeneutic argument that the exercise of power always happens within and is shaped by a moral framework. Ironically, Foucault exhibited a type of "view from nowhere," an inability to admit his own moral position. To ignore the moral realm, I thought, would lead his followers into a nihilism or at least a cynicism that could well lead to bitterness and hopelessness. In the title I called this the dead-end. I thought of us today, trying to make sense of our world, turning to the brilliant work of Heidegger and Foucault, but then finding ourselves in a dilemma caused by Heidegger's arrogance and Foucault's thinking that ends in a dead-end. To what or whom could we then turn? I suggested that Gadamer's appreciation of dialogue and his abhorrence of scientism could help us out of the dilemma, and I explained why.

That was a really fun article to write. One of the things I liked most about it is that there are aspects of Heidegger's thought that have similarities to interpretative Judaic traditions. It occurred to me that given how important Jewish life was in Germany before World War II, maybe Heidegger was affected by Jewish thought more than he could imagine. I suggested that pointing that out is maybe our best revenge.

At the same time that I was researching the Heidegger-Foucault dilemma, I was working on a different project, which was to take up once again the study of midrash, that is Jewish biblical commentary from the first millennium of the Common Era, which were stories about stories. I first studied (or tried to study) midrash when I was in the seminary, but at that time I really couldn't concentrate on it, because I was involved in the civil rights and anti-war movement, and at 23 years of age I was somewhat crazed and just not able to do the hard work that midrash required. But in 2002 I left my teaching job and my practice so that Karen and I could move back to the Northwest, I was teaching as an adjunct at various schools, and so I had some time to devote to a new project. For several years Frank Richardson had been pushing me to become more articulate about the sources of my moral and political commitments. And as the years went by, his arguments began to make sense to me. I came to realize he was right—we can't just assume that our commitments are common sense or self-evident or simply the product of rational inquiry. It is our responsibility to seriously learn about and explore the traditions that have constituted us, and then critique them in light of new events, experiences, ideas, political alignments. So I spent the years after 2003 going back to the texts of my first midrash class. I tried to really delve into the subject. This time around, I was able to apply myself better, plus after 35 more years of living and 26 years of doing therapy, I was a bit better able to appreciate the nuances of textual study and better understand the importance of midrashic process and its relationship to therapy.

In the process of studying midrash I learned more about historicity, interpretive processes, relationality, and anti-idolatry. I began to see connections between hermeneutics and midrash and psychotherapy. These were exciting days, meaningful days, as I realized that some of the disparate, far-flung aspects of my life had more harmony or coherence than I had realized or at least had been able to put into words. So in the following years I wrote about those connections, and the ideas I developed meant a great deal to me. I came to understand why certain political commitments of mine were so important to me. I learned more about what matters to me and why it matters so much. And I learned why I got involved in psychotherapy and why I do therapy in the ways I do and not in other ways. It turns out the hermeneuticists are correct: moral understandings are primary, they set the frame, and it is within that frame that a certain kind of politics is brought to light. In 2007 the article "A Burning World, An Absent God: Midrash, Hermeneutics, and Relational Psychoanalysis" was published, as was a chapter in the edited volume *Midrash & Medicine*, titled "The Danger of Cure, the Value of Healing: Toward a Midrashic Way of Being," in 2011. I'm proud of them; my life makes more sense to me now. Incidentally, *Midrash & Medicine* was edited by my old advisor at the seminary, Rabbi William Cutter, who taught the midrash class I was enrolled in, all those years ago. Sometimes, life comes full circle.

HM: I think you've answered this question, both in your earlier address here at the conference, and also in your most recent 2015(c) article in the journal *Comtemporary Psychoanalysis*, "Relational Psychoanalysis as Political Resistance," but what more do you think needs to happen for the field to right its course?

PC: "Relational Psychoanalysis as Political Resistance" was a long time in coming. I think I first gave a talk in Seattle about this topic in 2009. Then I expanded it in an invited paper I gave at Simon Frasier University in Vancouver, B.C. in 2013. It's gotten better over the years. It's a hermeneutic treatment of relational psychoanalysis. I kept trying to interest one of my dissertation students in writing a cultural history of relational psychoanalysis, but nobody would take it on, so finally I guess I just went and did it myself in an abbreviated form. The article is not as encompassing as a dissertation would be, which is too bad, but I gave it a try. What I'm trying to say is that we can understand Relational Psychoanalysis as a social site, a space in which a certain kind of lefty political resistance shows up. I don't mean that it's a wonderfully political movement. I'm arguing that in our social world, effective and sustained political action, especially on the left, is difficult to find, and the hegemony of mainstream, capitalist, consumer life and neoliberalism is so strong that it's difficult to know where to stand in order to contribute. The middle classes have certainly retreated from more overt activism. A lot of prominent relational analysts, wonderful folks like Lew Aaron, Neil Altman, Jessica Benjamin, Darlene Ehrenberg, Nancy Hollander, Lynne Layton, Steven Mitchell, Steve Seligman, Donnel Stern (the list goes on), were all activists to various degrees, lefties in the'60s. And yet only a few of them (Altman, Hollander, Layton, and younger ones such as Steve Botticcelli, Orna Guralnik, Eyal Rozmarin) now write about and bring into the therapy hour a more explicit politics. So where did that overt activism go?

In the article I suggest that resistance is still a force in the work of the relationalists, but in a much subtler way, and given the politics of health care, out of necessity a more disguised way. Helping patients develop a way of being that is more related, more compassionate, more critically discerning, more self-reflective, respectful of and more willing to learn from difference, are ways of helping others prepare themselves for a more explicit lefty activism. I also suggest that a common political critique of

Relational Psychoanalysis, that it is still too focused on the dyad and the isolated, rarified clinical moment, remains valid. I hope Relational analysts could find ways of being more directly, explicitly active politically, inside and outside the consulting room. But even as it stands now, relational practices still have political implications (See Cushman, 1998, 2012b/c).

Also, recently I wrote a chapter for a book titled *Rethinking Therapeutic Culture*. It's an unusual book about therapy, because the contributors are in American history, sociology, feminist studies. I'm the only therapist in it. Initially my title was "Living in the Politics of Uncertainty," although sadly the publisher (University of Chicago Press) retitled it "Practice." This was published in 2015(b). It was an attempt on my part to write about what it was like to be a therapist in the 1980s and 1990s with a full-time practice, and yet knowing something about what is called the critique of therapeutic culture. I tried to write about how difficult it was to know (and agree with) the critique and yet continue to practice.

I started off with a story about driving to my peer consultation group. Off to the right are the Berkeley Hills, and I look up as I'm driving and I see all of these big, beautiful, expensive homes and I think about some of my therapist friends who used to be lefty activists who are living in these homes. I said out loud to myself, "what are we doing? What are we *doing*?" And I'm thinking that we don't realize the political effects that we're having on the society.

We mistakenly think therapy is this isolated, encapsulated event that happens in the office. But we are exercising a certain amount of power in this society, and yet usually we don't look at it or examine it politically. So we don't know whether or to what extent our practices fit with or work against the political status quo.

That's partly what I was trying to write about explicitly in the political resistance article. How can we interpret the politics of Relational Psychoanalysis and how do we understand it politically even though almost nobody talks about it politically? There is a little bit of that these days, but not much. We could, for instance, interpret Jessica Benjamin's work, especially her idea about the doer/done to dynamic, as a disguised moral argument. So, the origins of my resistance article go back to the ideas Frank Richardson and Blaine Fowers and I started trying to put into words all those years ago at APA conferences in the early 1990s, like the symposium we organized in 1993 that Brent Slife published in the *Journal of Theoretical and Philosophical Psychology* as a special issue. My paper was titled "Psychotherapy as Moral Discourse" published in 1993. It was from that that paper that my more recent projects, including this most recent article, developed.

There is much that Relational Psychoanalysis delivers as part of a disguised moral position that inevitably has political meanings. But they're not explicitly drawn out, and I imagine that many analysts, even relational analysts, would be nervous about that idea. But I decided to write about it and see what happens (See Cushman, 2009b).

What psychologists stop ourselves from doing—and yet what we must do to fulfill our responsibilities to the public—is address ourselves to the political realm, especially in our role as psychologists, although we don't know how to do that. Psychologists often talk about issues that are implicitly political but we tend to reduce them to an acontextual scientism, especially in psychotherapy (See Cushman, 1991a, 1994, 2000b, 2013). It's maddening once you start thinking about this and realize the serious harm it can do and the important possibilities that are lost. I think we have a responsibility to the public to be truthful to them, to talk about this more explicitly and to be more active politically in a direct way. That's what I hope we will work on in the future.

Part of that political activism—and this will be the last thing, I promise—has to be criticism of how we are presenting ourselves to the public. That is what I hope the new book I'm trying to write is going to be about. The lies we tell the insurance company in order to get paid and the lies we tell the public about psychology as a STEM discipline and our so-called "evidence-based" practices, these have got to stop. They won't, of course, but they must stop and we must speak out about this.

The working title of my new book is *Psychotherapy in an Age of Illusion*. I want to write about living in an online world, because illusions are what electronic machines produce. Computers accomplish illusion by being procedural, which is what psychology has now become. Any of us who teach in psychology know that graduate programs are now being framed by competencies, which is an extreme form of proceduralism. Behavioral competencies cannot begin to capture the wild beauty of true learning. It is wrong to present these electronic grids of competencies as a representation of graduate learning. And yet, schools desperate to get or maintain APA accreditation have complied with this approach and present it to the public as if it is a great advance in scientific education. It is not a great advance, it is simply a reflection of our larger social world; it is mechanistic, instrumental, technicist drivel—it is procedural illusion.

I think we should really be ashamed of presenting competencies as learning and manuals as therapy. So, that's what I want to keep writing about and encouraging all of us to think about. We cannot pretend that our work has no moral framework. And we can't pretend it doesn't have political consequences. The paper I gave at this most recent Psychology and the Other conference (2015)—"The Golem Must Live, the Golem Must Die"—addresses these issues, especially the attempt of psychology to gain legitimacy and influence by colluding with insurance corporations and the Department of Defense's program of torture. The two topics have their origins in the same problem: the modern-era scientism that preaches the truth can only be found by bracketing off all prejudgments, thereby achieving a putative objectivity. The philosophical incoherence of that belief has been demonstrated many times in philosophy. It is that attitude—that objectivity guarantees the one truth—that made it possible for psychologists to claim that the use of evidence-based practices, academic competencies, and manualized therapies ensures that their practices are properly scientific or to justify their involvement in torture. Ethics becomes something one follows by rote, without the ability to engage in nuanced moral discourse. Ethics is something to be feared, something to be followed to avoid trouble with the law, not a process or a way of life one is continually involved in in order to live a good life. As long as moral discourse is thought to be the purview of other disciplines, psychologists—in their urgency to find legitimacy, influence, and money—will continue to risk being involved in shady or explicitly unethical acts.

HM: In the beginning of this interview we started with the idea of your shock at the invitation to conduct this interview but despite this we are honored you agreed to it and cherish your contributions to the field. Frank Richardson, who is with us here today, suggested that your paper on the "The Empty Self" actually helped revitalize Division 24 and has felt that your work, which speaks to deep human dynamics in wider cultural and social and historical contexts, has been a model for so many scholars within and beyond Division 24. He also suggested that your more recent work also continues to "set the tone" for how we might deepen our cultural critique, such as Jeff Sugarman's recent work on neo-liberalism, and maintain our intellectual humility at the same time. So, thank you very much.

PC: Thank you all for coming, and thank God it's over!

Bibliography

Cushman, P. (1981). Networking: A systems delivery approach to the treatment of cult membership. In D. A. Halperin (Ed.), *Psychodynamic Perspectives on Religion, Sect, and Cult* (pp. 343–352). Boston, MA: John Wright.

Cushman, P. (1984). The politics of vulnerability: Youth in religious cults. *Psychohistory Review, 12*, 5–17.

Cushman, P. (1986). The self besieged: Recruitment-indoctrination processes in restrictive groups. *Journal for the Theory of Social Behaviour, 16*, 1–32.

Cushman, P. (1987). History, psychology, and the abyss. *Psychohistory Review, 15*, 29–45.

Cushman, P. (1989). Iron fists/velvet gloves: Recruitment-indoctrination processes in a mass marathon psychology training program. *Psychotherapy, 26*, 23–29.

Cushman, P. (1990). Why the self is empty: Toward a historically situated psychology. *American Psychologist, 45*, 599–611.

Cushman, P. (1991a). Ideology obscured: Political uses of the self in Daniel Stern's infant. *American Psychologist, 46*, 206–219.

Cushman, P. (1991b). No empty fixes. *American Psychologist, 46*, 542–543.

Cushman, P. (1991c). Psychologic or psychological Esperanto?. *Psychological Inquiry, 2*, 339–342.

Cushman, P. (1992a). Psychotherapy to 1992: A historically situated interpretation. In D. Freedheim et al. (Eds.), *History of psychotherapy: A century of change* (pp. 21–64). Washington, DC: American Psychological Association.

Cushman, P. (1992b). Revealed, concealed, congealed, curdled: A reply to Pomichalek, Lamb, and Denner. *American Psychologist, 47*, 82–84.

Cushman, P. (1993). Psychotherapy as moral discourse. *Journal of Theoretical and Philosophical Psychology, 13*, 103–113.

Cushman, P. (1994). Confronting Sullivan's spider: Hermeneutics and the politics of therapy. *Contemporary Psychoanalysis, 30*, 800–844.

Cushman, P. (1995a). *Constructing the self, constructing America: A cultural history of psychotherapy*. New York, NY: Addison-Wesley.

Cushman, P. (1995b). Hermeneutics, history, and the present psychoanalytic moment. *fort da, Journal of Northern California Society for Psychoanalytic Psychology, 1*, 20–22.

Cushman, P. (1995c). On tapestries and entanglements: A response to commentaries on *Constructing the Self, Constructing America. Psychohistory Review, 24*, 77–98.

Cushman, P. (1996a). Disputed traditions: A review of Jane Flax's *Disputed Subjects. Psychoanalytic Dialogues, 6*, 477–488.

Cushman, P. (1996b). Locating dialogue: A reply to Flax. *Psychoanalytic Dialogues, 6*, 510–520..

Cushman, P. (1996c). More surprises, less certainty: Commentary on Alan Roland's "The influence of culture on the self and selfobject relations." *Psychoanalytic Dialogues, 6*, 277–288.

Cushman, P. (1998). On knowing when to get off the train: Communism, psychotherapy, and Nathan Adler's critical legacy. *The Other: The Journal of the Lacanian School of Psychoanalysis, 2*, 25–35.

Cushman, P. (1999). From emptiness to multiplicity: The self at the year 2000. *Psychohistory Review, 27*, 15–31.

Cushman, P. (2000a). White guilt, political activity, and the analyst. *Psychoanalytic Dialogues, 10*, 607–618.

Cushman, P. (2000b). Will managed care change our way of being?. *American Psychologist, 55*, 985–996.

Cushman, P. (2002). How psychology erodes personhood. *Journal of Theoretical and Philosophical Psychology, 22*, 103–113.

Cushman, P. (2004, October 19). The politics of consumption [guest editorial]. *Seattle Times*.

Cushman, P. (2005a). Between arrogance and a dead-end: Psychoanalysis and the Heidegger/Foucault dilemma. *Contemporary Psychoanalysis, 41*, 399–417.
Cushman, P. (2005b). Clinical applications: A response to Layton. *Contemporary Psychoanalysis, 41*, 431–435.
Cushman, P. (2007a). A burning world, an absent God: Midrash, hermeneutics, and relational psychoanalysis. *Contemporary Psychoanalysis, 43*, 47–88.
Cushman, P. (2007b). Response to Lewis Aron. *Contemporary Psychoanalysis, 43*, 113–120.
Cushman, P. (2007, Fall). Panel member at Thinking Critically in the Midst of the Maelstrom: Can Psychoanalysis Help Us Stay Sane in an Insane World?, a conference co-sponsored by Psychoanalysts for Social Responsibility and PsyBC.
Cushman, P. (2009a). A burning world, an absent God: Midrash, hermeneutics, and relational psychoanalysis. In L. Aron & L. Henik (Eds.), *Answering a question with a question: Contemporary psychoanalysis and Jewish thought.* New York, NY: Academic Studies Press.
Cushman, P. (2009b). Empathy—what one hand giveth, the other taketh away. *Psychoanalytic Dialogues*, 9, *121–137.*
Cushman, P. (2011). So who's asking? Politics, hermeneutics, and individuality. In R. Frie & W. Coburn (Eds.), *The challenge of individuality: New perspectives in psychoanalytic contexualism* (pp. 21–40). New York, NY: Routledge.
Cushman, P. (2012a). The danger of cure, the value of healing. In W. Cutter (Ed.), *Midrash & medicine: Healing the body and soul in the Jewish interpretive tradition* (pp. 211–233). Woodstock, VT: Jewish Lights.
Cushman, P. (2012b). Defenseless in the face of the status quo: Psychology without a critical humanities. *Humanistic Psychologist, 40*, 262–269.
Cushman, P. (2012c). Why resist? Politics, psychoanalysis, and the interpretive turn: A review of With Culture in Mind. *DIVISION/Review, 4*, 11–13.
Cushman, P. (2013). Because the rock will not read the article: A discussion of Jeremy Safran's critique of Irwin Hoffman's "Doublethinking Our Way to Scientific Legitimacy." *Psychoanalytic Dialogues, 23*, 211–224.
Cushman, P. (2015a). Book review of *Contemporary Psychoanalysis and the Legacy of the Third Reich.*. Contemporary Psychoanalysis, *51*, 176–184.
Cushman, P. (2015b). Practice. In T. Travis & T. Abrey (Eds.), *Rethinking therapeutic culture.* Chicago, IL: University of Chicago Press.
Cushman, P. (2015c). Relational psychoanalysis as political resistance. *Contemporary Psychoanalysis, 51*, 423–459.

Cushman, P. (2017). The golem must live, the golem must die: On the moral imperative to write critical cultural histories of psychology. In D. Goodman & E. Severson (Eds.) *Memories and Monsters: Psychology, Trauma, & Memories.* New York: Routledge.

Cushman, P. (in press). Abelove's historical vision: Gentle warnings against moralism, colonialism, and an apolitical identity politics. *Studies in Gender and Sexuality.*

Foucault, M. (1970). *The order of things: An archaeology of the human sciences.* New York, NY: Vintage Books.

Foucault, M. (1973). *The birth of the clinic and archaeology of medical perception.* London: Tavistock.

Friedlander, S. (1993). *Memory, history and the extermination of the Jews of Europe.* Indianapolis, Indiana: Indiana University Press.

Gadamer, H. G. (1989). *Truth and method* (2nd rev. ed.). New York, NY: Crossroads. (Original work published 1960)

Gilford, P., & Cushman, P. (2001). Wishing away history: A response to Hoyt. *American Psychologist, 56,* 765–766.

Kohut, H. (1971). *The analysis of the self: A systematic approach to the psychoanalytic treatment of narcissistic personality disorders.* New York, NY: International Universities Press.

Lowe, D. (1983). *History of bourgeois perception.* Chicago Il: University of Chicago Press.

Sampson, E. (1981). Cognitive psychology as ideology. *American Psychologist, 36*(7), 730–743.

Sampson, E. (1988). The debate on individualism: Indigenous psychologies of the individual and their role in personal and societal functioning. *American Psychologist, 43*(1), 15–22.

Sluga, H. (1993). *The Heidegger crisis: Philosophy and politics in Nazi Germany.* Cambridge, MA: Harvard University Press.

Stern, D. (1990). Courting surprise: Unbidden perceptions in clinical practice. *Contemporary Psychoanalysis, 6*(3), 452–478.

Wolin, R. (1993). *The Heidegger controversy: A critical reader.* Cambridge, MA: MIT press.

Heather Macdonald is Assistant Professor of Psychology at Lesley University. She came to academia after years of practice as a clinical psychologist whose work involved community outreach, child assessment, and individual therapeutic services to children and families in the foster care system and with youth involved in the juvenile justice system. Her work in urban environments and abroad has led to scholarly research on the interface between culture, social justice, relational ethics, clinical practice, and postcolonial thought. Her first monograph, titled *Cultural and Critical Explorations in Community Psychology* (2017) published by Palgrave Macmillan, further considers the implications of psychological assessment, diagnosis, and historical trauma. As much as the book represents a cultural critique of more traditional clinical discourse, it also suggests that if we view clinical relationships from the perspective of a relational ethics, that includes discourse about cultural memory, freedom, power, and other events that break with the neoliberal expectation of "mutual reciprocity," then therapeutic work can also be a site of transformation and reversal that moves beyond fixed racial categories. She is a fellow of the Psychology and the Other Institute.

A Phenomenological-Contextualist Perspective in Psychoanalysis: A Conversation with Robert D. Stolorow

(Interviewed by Peter N. Maduro, Private Practice)

In the interview that follows, Robert D. Stolorow, Ph.D., Ph.D., embodies the phenomenological-contextualist principles that undergird his and his collaborators' psychoanalytic framework. This framework, which they have named intersubjective-systems theory, is phenomenological, writes Stolorow elsewhere (2011, p. 19), "in that it investigates and illuminates worlds of emotional experience", and it is "contextual in that it holds that such organizations of emotional experience take form, both developmentally and in the psychoanalytic situation, in constitutive intersubjective contexts" (Stolorow 2011, p. 19). As is self-evident in those formulations, and laid bare to the reader in this interview, of paramount *substantive* importance in Stolorow's clinical theory—which he describes as a form of "applied philosophy"—are emotional life and the relational contexts that are constitutive in whether and how it is experienced. And of paramount *methodological* importance is empathically and introspectively informed

R.D. Stolorow (✉)
Institute of Contemporary Psychoanalysis, Santa Monica, California, USA
e-mail: robertdstolorow@gmail.com

H. Macdonald et al. (eds.), *Dialogues at the Edge of American Psychological Discourse*, Palgrave Studies in the Theory and History of Psychology, DOI 10.1057/978-1-137-59096-1_5

in-depth inquiry into, and contextual understanding of, the individual person's distinctively organized subjective world. In addition, implicit in Dr. Stolorow's brand of in-depth inquiry is an attitude of respect for the personal and particular of any individual's unique experience as well as for its rich, constitutive contextuality.

Reflecting these values in spades, in the below interview Dr. Stolorow shares a historically-oriented, in-depth, and nuanced account of the development of the theoretical convictions at the core of his intersubjective-systems perspective. He elucidates in no uncertain terms how this development was significantly constituted by the personal, educational, and professional relationships with important others (many of them "nominally psychologists"), the scientific tensions within the academy (mainstream psychology versus Henry Murray's (1938/2007) "personology"), certain timing-factors (e.g., educated at Harvard University when clinical psychology was organized as an interdisciplinary program and instructed by icons in their fields), and other fortuitous or un-fortuitous circumstances (including personal trauma) that were its context. Moreover, his account conveys his attitude of profound respect and care for the particulars and contextual embeddedness of his own, and his theory's, historical truth.

In these ways, Dr. Stolorow's interview is reminiscent of an in-depth case study in a treatment context: it is an analogue to a respectfully careful inquiry into, and illumination of, the nuances of both a person's emotional phenomenology and the relational-contexts in which it takes form. His historical distillation of the central tenants of his intersubjective-systems theory is akin to an analytic illumination of the emotional convictions—saturated as they necessarily are in the lived-contexts from which they emerge—that together constitute a person's "character." In short, we see in this interview that just as emotional life takes form, goes on, and transforms in its constitutive relational contexts, so too has Dr. Stolorow's clinical sensibility and theory taken form and transformed in multi-faceted relational context.

Of special note, at least to this interviewer, is Dr. Stolorow's insistence on the context of personal trauma—and defense against trauma, in the form of "metaphysical illusion"—in theory-formation. As he and George Atwood reflect elsewhere (2016, p. 292), "[a]gain and again we have been led to the inseparability of theoretical thought and the life in which it emerges." In the interview, Dr. Stolorow courageously stands behind this contextualist conviction by avowing the impact of trauma on his own theoretical-self, as well as by critically disclosing its possible impact in others' psychoanalytic theorizing.

Dr. Stolorow's erudition is co-extensive with his extensive professional accomplishments. He is a Founding Faculty Member and Training and Supervising Analyst at the Institute of Contemporary Psychoanalysis, West Los Angeles; a Founding Faculty Member at the Institute for the Psychoanalytic Study of Subjectivity, New York City; and a Clinical Professor of Psychiatry at the UCLA School of Medicine. He received his Ph.D. in Clinical Psychology from Harvard University in 1970 and his Certificate in Psychoanalysis and Psychotherapy from the Psychoanalytic Institute of the Postgraduate Center for Mental Health, New York City, in 1974. He also received a Ph.D. in Philosophy from the University of California at Riverside in 2007. He holds diplomas both in Clinical Psychology and in Psychoanalysis from the American Board of Professional Psychology (ABPP), and he is a Fellow in the Divisions of Psychoanalysis and Humanistic Psychology of the American Psychological Association. He received the Distinguished Scientific Award from the Division of Psychoanalysis in 1995, the Haskell Norman Prize for Excellence in Psychoanalysis from the San Francisco Center for Psychoanalysis in 2011, and the Hans W. Loewald Memorial Award from the International Forum for Psychoanalytic Education in 2012.

As the more detailed list of references below indicates, Dr. Stolorow is the author of *World, Affectivity, Trauma: Heidegger and Post-Cartesian Psychoanalysis* (2011) and *Trauma and Human Existence: Autobiographical, Psychoanalytic, and Philosophical Reflections* (2007), and coauthor of *Worlds of Experience: Interweaving Philosophical and Clinical Dimensions in Psychoanalysis* (2002), *Working Intersubjectively: Contextualism in Psychoanalytic Practice* (1997), *Contexts of Being: The Intersubjective Foundations of Psychological Life* (1992), *Psychoanalytic Treatment: An Intersubjective Approach* (1987), *Structures of Subjectivity: Explorations in Psychoanalytic Phenomenology and Contextualism* (2014[1984]), *Psychoanalysis of Developmental Arrests: Theory and Treatment* (1980), and *Faces in a Cloud: Intersubjectivity in*

Personality Theory (1993 [1979], 2nd. ed.). He is also coeditor of *The Intersubjective Perspective* (1994) and has authored or coauthored more than two hundred articles.

At the close of the interview, Dr. Stolorow offers to pass the "baton" of critical psychoanalytic contextualism on to those of us influenced by his work. Perhaps readers of the below interview will join me in the hope that, whatever the rest of us do, Dr. Stolorow keeps running with the baton.

Peter N. Maduro (PM): Now in autobiographical accounts you've offered here and there, including recently in your and George Atwood's (1984) articles, "Legacies of a Golden Age: A Memoir of Collaboration," which also appears in the second edition of your third book, *Structures of Subjectivity,* and the 2016 article, "The Phenomenological Circle and the Unity of Life and Thought," you've credited many who were nominally psychologists with profound influence on you. Among them were Rollo May, Robert White, Henry Murray, Silvan Tompkins (whom I might add you describe as "one of the great theorists of 20th century psychology") and of course George Atwood.

Further, after exploring studies in philosophy at the undergraduate level, and a brief five-week stint in medical school with an eye toward doing "hard science research in psychopathology," you ultimately pursued a doctorate in clinical psychology at Harvard. Then in 1972, after receiving your PhD in psychology and initiating psychoanalytic training in New York, you joined Tompkins and Atwood on the faculty of Livingston College at Rutgers University. So at least nominally you were an academic psychologist.

I have reviewed this slice of your history in order to establish that you have a lot of psychology in your personal and professional history, and yet, when you were originally invited to participate in this project, you said that you do not identify with the discipline of psychology. So I'd like to launch our conversation by asking you the following two questions. I'll state both of them up front.

First, why and in what ways do you not identify with the discipline of psychology? I was thinking that you might take an historical approach, as you've done in other settings, to your response to this question, and in so doing discuss the origins and development of your intersubjective-systems theory, the personal and intellectual contexts in which it emerged, as well as those frameworks for understanding psychological life from which it rebels and offers an alternative.

And the second question, at least for purposes of the balance of this interview, is this: how might one most accurately characterize the field of inquiry and therapy with which you do identify?

Robert Stolorow (RS): I think the key word in what you said when you described my association with other people who hold degrees in psychology was the word *nominally*, "nominally psychologists." But I'll get back to that in a moment.

As you indicated, the first studies that I undertook after finishing my undergraduate work, which was at Harvard, was to go to medical school. As an undergraduate I had a very strong background in the hard sciences, maybe difficult for you to believe, but I majored in mathematics for two and a half years, and then physics and then biochemistry and biology. And my idea was to go to medical school because I thought it served as the best background for doing what I wanted to do, which was hard-science research in severe psychopathology. That's what I wanted to do.

However, it didn't take long for me to realize that medical school was a terrible fit for me. The psychiatry courses in first year medical school were insubstantial compared to what I had already done in taking a couple of psychology courses as an undergraduate. Anything involving eye-hand coordination I was terrible at, such as anatomy lab. So, I like to say, many lives have been saved by my decision to drop out of medical school after five weeks; a decision that took the scalpel out of my hands.

I decided that a better background for what my ambition was at the time was to do a doctorate in clinical psychology. I got accepted to Harvard again, and that's where I decided to go.

Now, at the time when I was just beginning graduate school, I was still under the spell of the kool-aid of the scientist-practitioner model of clinical psychology. I had drunk that kool-aid completely. I was still wedded to the idea of really being a scientist. However, when I got to Harvard I was pleasantly surprised to find other elements in the training there that had very little to do with being a scientist, at least as I conceived of being a scientist.

For one thing, the clinical psychology program at Harvard was not located in a Psychology Department, thankfully. It was located in the Department of Social Relations, which was a department created by scholars from four disciplines—personality psychology, social psychology, sociology, and cultural anthropology.

PM: So the Psychology Department at Harvard at the time was structured as an interdisciplinary field?

RS: No, not exactly. The *clinical program* was not in a Psychology Department. There was a Psychology Department, but it was dominated by behaviorists. For example, B.F. Skinner, as I recall, was in it.

So, thankfully, the clinical psychology program was in this interdisciplinary department. And by the way, the founding scholars from the four disciplines were all interested in psychoanalysis, and they all had been analyzed. That was quite a group! So instead of having to study the experimental psychology of rats, for example, I got to take sociology from Talcott Parsons, cultural anthropology from John Whiting, and identity formation from Erik Erikson. It was an extremely rich background and a really very wonderful fit for my nascent contextualist sensibilities, which I'll get to later.

PM: From the sounds of it, indeed it was!

RS: Yeah. So that was one thing. The other thing was that the clinical psychology program at that time was the first, and sadly the last,

> stronghold of a tradition in academic personality psychology known as *personology*. This was a perspective created by Henry Murray at the Harvard Psychological Clinic. Its basic premise was that knowledge of human personality was to be advanced by the systematic in-depth study of the single individual, the systematic in-depth case study.

Now as I'm sure you're aware, such a position is totally at odds with the philosophy of science that has dominated the academic discipline of psychology—back then and to this day. So the actual atmosphere that I was exposed to was very much contrary to the view of hard-science psychological research that I had when I entered the program.

What happened is that over the course of my first two years of studies there I became completely disillusioned with hard-science psychological research. From actually being exposed to it and being involved in it, I learned that the process of identifying variables, quantifying them, measuring them, doing statistical analysis of these quantified variables, and so on, strips virtually everything humanly meaningful from the study being conducted. I became completely disillusioned with that and I played with the idea of doing a second doctorate in philosophy concurrently with finishing my doctorate in clinical psychology.

During the year in between graduating from college and starting graduate school at Harvard I took a course from Rollo May, as you mentioned, at the New School for Social Research in New York, a course in existential psychology, and got exposed to the ideas in phenomenology—Husserl, Heidegger, Ludwig Binswanger, and so on, and I was fascinated by these ideas in phenomenology and existential philosophy and psychology.

In my second year of graduate school, I contacted a former philosophy professor of mine, Henry Aiken, who had moved from Harvard to Brandeis, and presented him with the idea that I wanted to use philosophy to clean up the mess that was psychoanalytic theory—this was the influence of the Rollo May course—and to pursue a second degree in philosophy.

Aiken was very supportive of my idea, although it turned out that the faculty there at Brandeis would have had me come as a post-doc rather than doing a second doctorate. But in the meantime, in my third year, I did a clinical internship at a place that was very psychoanalytically oriented and that offered terrific psychoanalytic supervision, the Massachusetts Mental Health Center, and I found that I really enjoyed psychoanalytic work. So instead of going to Brandeis for a post-doc in philosophy, I went to New York to pursue psychoanalytic training. However, my love of philosophy remained with me, even though pursuing it rigorously was going to have to await more than three decades after finishing my doctorate in psychology.

I used to say that I was married to psychoanalysis, but that philosophy was my mistress—until 2007, when I consummated my marriage to philosophy as well and became a bigamist. Maybe that's my identity, maybe that's what I identify with.

PM: Bigamy? Ha, yes, you're a bigamist married to both philosophy and psychoanalysis.

RS: Yeah, an interdisciplinary bigamist. Now to round out the story a little bit more, in 1972 in the second year of my analytical training, I decided that I wanted to pursue an academic career, and I learned of a position opening at Rutgers, as you mentioned, where George Atwood and Silvan Tomkins were on the faculty. It turned out that there was a group there, at Livingston College of Rutgers University, that was interested in resurrecting personology at Livingston College—resurrecting the in-depth case study. I remember that, after hearing a presentation I gave there as part of the application process, Silvan Tomkins called me up and said that if I came to Livingston College we would have a "critical mass," as he put it, for starting up a new doctoral program in personologically-oriented personality psychology.

There were actually several meetings devoted to that vision, but it never got off the ground, largely because of Silvan. Silvan was the only senior person in this group who had any clout, and he fell into a kind of a depressive mood that stood in the way of his active involvement in the project. So it kind of fell by the wayside.

However, there was one important result that came out of this project. Before going further with that I should mention that, although the people in this group were nominally psychologists, they didn't look like any psychologist in the mainstream of the discipline. They were all personology people, which was very much on the periphery of academic psychology and has remained so.

PM: Just to be clear, if, as some say, a domain of inquiry is significantly delineated and defined by its methodology, then would we say the discipline of "personology" was defined by the in-depth case study of the individual person and his or her life-world, as compared to the variables, quantifications, double-blind studies, and so forth, of conventional scientific empiricism?

RS: Exactly. So really the so-called scientist-practitioner model, in the sense of the prevailing notion of what hard science is supposed to look like, was nowhere to be found in personology. It was what nowadays is called *qualitative research*. So yes, this was a group of nominal psychologists, but not really in the mainstream of psychology. A bunch of outsiders, which was very attractive to me at the time because I was already very much disaffected with the mainstream of psychology.

PM: Incidentally, was there a more mainstream research psychology sphere within Rutgers's departmental structure, like there was at Harvard, which was separate from where you, Silvan Tomkins, and George Atwood were located?

RS: It was informally separated, not formally. Rutgers was a university that was made up of a number of separate colleges and this program that we were trying to establish was in Livingston College. The mainstream psychology people, at least in clinical psychology, were in Douglas College, if I'm remembering correctly.

In any case, the personology project failed largely because of Silvan's collapsed mood. But the one thing that came out of it, the one concrete thing, was a series of personological studies conducted by George Atwood and myself during the early and mid-1970s, of the personal subjective origins of four psychoanalytic theories—Freud's, Jung's, Wilhelm Reich's and Otto Rank's—in which we used the Murray (1938/2007) methodology of the systematic in-depth case study to investigate each theorist's life and work. Looking at the parallels and unities in the thematic structures of the works, on the one hand, and the lives, on the other hand, we found in each case that the theories were significantly shaped by the personal experiential world of the theorist.

In 1976 we gathered these studies together into a book called *Faces in a Cloud: Subjectivity in Personality Theory*. By the way, George and I referred to the first edition of this book as *the bleeder*, because it was so cheaply done that the dye from the book actually came off on your hands.

PM: I remember hearing about this. The book-cover was red, wasn't it?

RS: It was red, that's right. So we called the book *the bleeder*.

PM: Yes, two years ago I was handling and perusing one of these original hard-copies in your friend Shelley Doctors's office in New York City. She warned me then that it had bled red dye in her hands when she first read it years before!

RS: In the concluding chapter of the book we reasoned as follows: Since psychoanalytic theories can be shown to a significant degree to be expressions of the subjective world of the theorist, what psychoanalysis needed was a theory of subjectivity itself. That is, a theory broad enough, and inclusive enough, to encompass not only the phenomena that the various other theories addressed, but also the theories themselves as psychological products. We also proposed that what this theory should be is a theory of subjectivity itself. This was George's and my first step in our lifelong project of recasting psychoanalysis as a form of phenomenological inquiry, a form of inquiry concerned with investigating worlds of emotional experience, the structures that prereflectively organize them, and the intersubjective contexts in which these structures take form.

The sentence that I just spoke would be a way of characterizing our psychoanalytic perspective. But is this psychology? No, it's not really psychology; it wouldn't really be recognized as psychology in any traditional sense by most people. I think what it would be seen as, and what it in fact *is*, is a form of applied philosophy. It's a form of phenomenology aimed not at the universal structures of experience, but at the particular structures of experience that develop in particular contexts in particular individuals. So that's how I would characterize it. I have concluded after a number of decades that psychoanalysis, in its essence, is applied philosophy.

PM: So your interdisciplinary bigamy produces a practice of applied philosophy? Is there any other adjective you might throw into that phrase "applied philosophy" that might accent how, instead of pure philosophy with its interest in the universal structures of human being and subjectivity, its focus is on the subjective structures that are the legacies of an individual person's experiences in the world? Is there any additional term that might zero in further on how applied philosophy aims to see and understand these lived particulars?

RS: I think the term that seems to best fit our perspective is *phenomenological contextualism.*

PM: Okay.

RS: Structures of experience in context.

PM: Right.

RS: So that was the beginning in 1976. Shall I continue with the historical unfolding?

PM: Yes, please. However, I'd like to build in one of my other questions for you at this point. Would you further address the ways your intersubjective framework was reactive to what you and George Atwood have from time to time called *metaphysical illusion*? I ask because in your aforementioned article, "The Phenomenological Circle. . . . ", I believe, you state that your effort to refashion psychoanalysis as a phenomenological contextualism "led you inexorably to a deconstructive critique of psychoanalytic metapsychology." So I'd love for you to comment on that as you proceed.

RS: Okay. Our dislike of these very abstract experience-distant formulations of various psychoanalytic metapsychologies was already present in the beginning and contributed to our wanting to refashion psychoanalysis as a form of phenomenological inquiry.

Going back to *Faces in a Cloud*, we found that the concepts that were most significantly shaped by the personal subjectivities of each theorist were the most abstract and experience-distant ones—the metapsychological concepts. In each case—and this was something that drove me crazy as a young analyst in training—you would have the co-presence of concepts that are far removed from clinical experience and utterly useless for learning how to do psychoanalytic work. Freudian drive theory, for example, coexisted confusedly with Freud's valuable clinical insights. George S. Klein in his book *Psychoanalytic Theory: An Exploration of Essentials*, also in 1976, proposed that Freud's theory actually amalgamated two theories—a metapsychology, which was an abstract experience-distant instance of the scientific materialism of the day, mixed with a clinical theory that was a hermeneutic theory of unconscious meaning.

So you've got these two frameworks coexisting that are really incompatible. And I remember so many times as a candidate in psychoanalysis thinking to myself, "Well, what the hell good is this for doing clinical work? And for understanding the experience of human beings?" So right from the beginning there was a kind of antipathy on our part toward these universalized absolutized experience-distant frameworks that have little to do with understanding emotional experience. But that antipathy became more and more systematic when I became more deeply immersed in philosophy and the critique of metaphysics.

That's getting ahead of myself because I want to stay in 1976 for a while.

PM: Okay, let's go back to that year and resume.

RS: There were several other important steps that we took in 1976. In addition to completing *Faces in a Cloud* and making our first initial proposals for a psychoanalytic phenomenology, I wrote an article in 1976 on the concept of psychic structure in which I suggested that we eliminate the metapsychological concepts of psychic structure, like id, ego, and superego, and instead think of psychological structure in terms of the principles or themes, the meaning structures, that organize emotional experience. It was in that article that I introduced the concept of *prereflective organizing principles* as the meaning of psychological structure that I was proposing. That conception became a very important idea in our evolving psychoanalytic phenomenology, including its importance for the concept of transference that I'll get to later.

A second thing that happened in 1976 was a meeting that George Atwood and I had at the Homestead Bar in New Brunswick, New Jersey, the hangout for Rutgers people, in which we were trying to put the finishing touches on an article on the phenomenology of the psychoanalytic situation. We were trying to think of a term that could capture—for the subtitle of a section on transference and countertransference—the impact on the analytic process of unconscious correspondences and discrepancies between the subjective worlds of patient and analyst. We thought of *interactional,* but that was too generic and nondescript. We thought of *interpersonal,* but we didn't like that because the interpersonalists at that time were much too focused on overt social behavior: who's doing what to whom.

Then one of us hit on the idea *intersubjective perspective.* We wanted a term that would capture the interplay of two subjective worlds, two worlds of subjective meaning. And one of us suggested *intersubjective*, and I don't remember which one of us—that's the one that stuck, because it seemed to capture perfectly the domain of inquiry that we were interested in and that we wanted to focus on. So that was when we introduced the term *intersubjective perspective.*

PM: So that formulation was born in 1976?

RS: That's 1976. These articles were published in 1978. *Faces in a Cloud,* even though we finished it in 1976, wasn't published until 1979. *Intersubjective perspective* was the term that stuck. Lew

Aron, a prominent relational theorist, credited that article with introducing the concept of the intersubjective perspective into American psychoanalytic discourse. So there in 1976 we already had the broad outlines of a phenomenological contextualism. We had the idea of rethinking psychoanalysis as a form of phenomenological inquiry. We had the concept of prereflective organizing principles. We had the contextual aspect captured by the term *intersubjective perspective*. So that was 1976, a good year for us!

PM: I am seeing better how your intersubjective perspective, or "phenomenological contextualism", took form progressively from very early on. One root clearly springs from your affinity for Murray's "personology" while at Harvard, and your related antipathy for the methods of mainstream psychology that you found stripped everything personal out of the study of personal experience.

RS: That's right. The next big step in that direction, particularly in the contextual direction, came about as a result of my meeting Bernard Brandchaft at a conference in 1979—speeding ahead on the axis of time here. Bernie gave a wonderful paper at this conference on so-called negative therapeutic reactions, the phenomenon that supposedly occurs when the analyst's correct interpretations make patients worse rather than better.

Bernie, in a very courageous paper at the time, maintained with clinical examples from his own practice that these adverse reactions occurred when the analyst's interpretive stance retraumatized the patient. That's why it made them worse. I had actually written a little section for the last chapter of a book I was doing with Frank Lachmann on developmental arrests that made a similar point. So I went up to Bernie afterward and talked about that. He asked me to give a paper at a conference he was chairing the following year, 1980, at UCLA on borderline pathology. I was kind of sick of that topic by then and said that I would do it, but only if he would collaborate with me, which he agreed to do.

In discussing the paper we realized that we had made very similar clinical observations, I from the east coast, New York, and he from the west coast, Los Angeles. Two people, different parts of the country, different ages. Bernie was about 25 years or more older than I was and we had different backgrounds. Bernie was an M.D. who went through psychiatry and traditional psychoanalytic training and so on. But nonetheless we had arrived at very similar observations, which were: If you take a very vulnerable archaically organized patient and put that patient in treatment with someone who worked according to the theory and technical recommendations of Otto Kernberg, very soon that patient will show all of the characteristics that Kernberg ascribes to borderline personality. The intense aggression, the emotional chaos, and so on, and the pages of Kernberg's books will come alive right before your eyes.

On the other hand, if you take that same vulnerable archaically organized patient and treat the patient according to the theory and technical recommendations given by Heinz Kohut, pretty soon that patient will look like a severe narcissistic personality disorder, and the pages of Kohut's books will come alive. Until there's a significant disruption in the transference, and then the patient will start looking like one of Kernberg's borderline patients, and the pages of Kernberg's books will come alive. So what this amounted to was that the manifest psychopathology, the manifest clinical picture, was not a product of the patient's psychological structures alone. It was codetermined by the patient's psychological organization and how this was understood and responded to by the analyst.

Further, having arrived at that formulation, Bernie, George, and I realized that it was a formulation that applied in general to all forms of manifest psychopathology, from the psychoneurotic to the overtly psychotic. That in no instance could the pathology—the manifest pathology—be understood solely in terms of the isolated mind of the patient. It was always codetermined in an intersubjective field to which the analyst's understanding, or lack thereof, was making a major contribution. We applied a similar idea to phenomena like intractable resistances, for example. So the contextualization of the phenomena of clinical psychoanalysis just took off at that point.

PM: These articles were a real challenge to what you have called the "doctrine of the isolated mind" that undergirded traditional psychoanalytic frameworks.

RS: Absolutely. There's a radical contextualization of everything psychoanalytic.

PM: So could we say that your collaboration with Brandchaft, and everything that followed from it, entailed another major step in your contextualist reformulations and critiques of "isolated mind" psychoanalysis—a step that included fairly candid exposures of how certain then contemporary psychoanalytic frameworks—like Kernberg's and aspects of Kohut's—risked decontextualizing psychological life in ways that would obscure clinical truth?

RS: That's right. This array of works contextualizing various clinical phenomenon took us into the mid-1980's and eventuated in the book *Psychoanalytic Treatment: An Intersubjective Approach* (1987).

There are two other early articles I would like to mention because they both played a very important role in the further development of the perspective. These were published in 1985, and both of them also became chapters in the book *Psychoanalytic Treatment: An Intersubjective Approach.* One was an article by me and Frank Lachmann on transference.

The chapter in the book has the title "Transference: The Organization of the Experience." In that chapter we criticized all the traditional concepts of transference—as displacement, projection, distortion, and so on. And instead proposed that transference in its essence consists in unconscious or prereflective organizing activity.

Now this idea was no accident because it links up transference with the concept of prereflective organizing principles, which I came to see as the basic building blocks of personality development. By the way, as an aside, what do we mean when we talk about a person's character or character structure? What we mean by that from a phenomenological perspective is the totality of prereflective organizing principles that shape a person's emotional experience. That's what we mean by character.

PM: As an aside I might just say I'm aware of an excellent blog essay of yours, entitled "What Is Character and How Does It Change?", that elaborates on this idea of character. For readers, your blog-home is at "Psychology Today On-Line" and it is called, "Feeling, Relating, Existing," right?

RS: Right, and a more extended version of that is going to appear in a philosophy book on character.

PM: Great.

RS: So the concept of transference as prereflective organizing activity unites with the view of psychological structure as prereflective organizing principles and with character as the totality of prereflective organizing principles that shape a person's experiences.

Now one advantage of thinking of transference as prereflective organizing activity is that it's inherently contextual, because you can't organize nothing. In order for something to be organized according to a prereflective organizing principle, there has to be *something* coming from the side of the analyst, for example, that is lending itself to being organized that way. So this conception of transference contextualizes it radically.

PM: And this is to say nothing of the fact that the organizing principles themselves are constitutively relational in so far as they are in part grounded in the patient's history of experiences with others.

RS: Exactly. So instead of seeing the patient's transference experiences as displacements or projections onto the analyst as a blank screen, which is an incredibly self-serving illusion—that any other human being can be a blank screen for another human being—instead of that you've got all kinds of complex intersubjective exchanges taking place between patient and analyst, with various things coming from the side of the analyst lending themselves to the particular organizing principles that the patient brings to the encounter.

Now we found it very useful to distinguish two broad classes of organizing principles, or to put it another way, two broad dimensions of transference. One we call the *developmental dimension,* in which the patient longs for the analyst to be a source of developmental experiences that were missed or lost or aborted during the formative years; the other we call the *repetitive dimension,* in which the patient anticipates, expects, fears, or actually experiences a repetition with the analyst of early developmental trauma—like the patients Bernie described in his paper on negative therapeutic reactions.

Each of these dimensions can be subdivided into multiple sub-dimensions, multiple developmental longings, multiple experiences of developmental trauma, and so on, leading to a multiplicity of organizing principles of both types. So both dimensions and their sub-dimensions are co-present in the therapeutic situation. But the therapeutic relationship tends to be dominated by one or another of them, depending on what's coming from the side of the therapist and how that is lending itself to one or another of these dimensions.

So you've got a picture of multiple dimensions of experience oscillating between the background and foreground of the patient's experience in response to the meanings of particular happenings within the intersubjective field. Now the same description also characterizes the analyst's transference, which is no different from the patient's—multiple dimensions of experience oscillating between the background and foreground of the analyst's experience in response to the meanings of particular goings-on in the intersubjective field.

So you've got an extremely complex picture here of two fluidly oscillating experiential worlds, each with multiple dimensions of experience oscillating between the background and foreground in response to the meanings for each participant of particular happenings in the intersubjective field, and each of these multidimensional fluidly oscillating experiential worlds mutually influencing one another. Such a complex picture is not appealing to those who are looking for solid bedrock to stand on. There is no bedrock to be found here; just phenomenological contextualism all the way down!

PM: I was just thinking how the notion of "therapist as blank screen" effectively suggests the therapist doesn't have any subjectivity that is part of the system. I could imagine a motivation within the "blank screen" theorist to de-complexity the relational field simply in order to make it more understandable. Maybc it'd be a product of the theorist's "complexity dread" or some such affect. But anyway I digress.

RS: Actually, I think that's a very good point. It alludes to one of the things that creates an aversive reaction to our perspective in some people. George and I referred to the basis for that aversion as *a fear of structureless chaos*. The picture of the fluidly oscillating complexity of the intersubjective field can easily evoke that fear of structureless chaos.

PM: You're alluding to a fear and aversion in some therapists for theory that doesn't provide clear answers?

RS: Particularly those people who need to feel that they're standing on solid bedrock. The philosopher Richard Bernstein referred to that fear as *Cartesian anxiety*.

PM: Why did he call it Cartesian?

RS: Because it's the opposite of what Descartes was searching for—namely, clear and distinct ideas.

PM: Among the emotional demands of your phenomenological-contextualist perspective, you're pointing to the way understanding the therapeutic situation in such complex, relational terms can evoke uncertainty anxiety, and thus require our tolerance of it in order to do thorough clinical work. In this regard, another feeling that I see in my own work, and in my supervision of candidates, is a kind of incompetence anxiety. I think it derives from holding clarity and distinctness as a personal, professional ideal since it sets the stage for feeling failure when what we see isn't so clear and distinct. I know I can feel this when I expect myself to clearly understand the complex clinical exchanges in front of me. One consequence of this feeling of failure is that it undermines my tolerance of clinical complexity and ambiguity. It can be very painful, especially when I, or the supervisee I'm working with, is already in a mood of self-doubt for one reason or another. So that's another thought I have, namely, the problem of incompetence anxiety or pain, and how it might inhibit one's openness to the complexities of psychoanalytic treatment.

RS: A good point, definitely.

The other article I wanted to mention, published in 1985 and reappearing in the book, *Psychoanalytic Treatment,* is one that I wrote in collaboration with my late wife, Daphne (Dede) Socarides Stolorow. This is an article entitled "Affects and Selfobjects" (Socarides & Stolorow, 1984/1985). To make a very long story short, this arose out of an attempt to revise Kohut's concept of selfobject in a way that would divorce it from drive theory once and for all and resituate it in affect theory. Without going into too much detail, Kohut's concept of selfobject referred to the need for certain forms of relationship in order to maintain or consolidate or restore the cohesion, continuity, and affective tone of one's sense of selfhood. That's how I would put it.

What Dede and I came up with was the idea that this phenomenon actually points to the centrality of affect and of integrating affect in consolidating the core of selfhood. That the core of the sense of selfhood is essentially integrated affect. And so we proposed that the essence of the selfobject concept was the need for attuned responsiveness to affect states in all phases of the life cycle; that the absence of such attuned responsiveness led to traumatic states, whereby affect could not be integrated and therefore was felt to be overwhelming and unbearable. Thus arises the need to defend against affect states.

So this was kind of a revolutionary idea. The article, as you might expect, was not particularly well-received by the orthodox Kohutians, because it didn't come from Kohut. I always wondered how Kohut might have responded to it had he been alive. But it actually proved very influential for clinicians throughout the world. People found it an extremely helpful way of thinking about so-called selfobject functions.

That led to a number of significant theoretical developments that were also of great clinical importance. For example, it led to a way of understanding so-called intrapsychic or psychological conflict—that such conflict comes about with regard to affect states that do not meet with attuned responsiveness from the surround, giving the child the sense that his or her emotional life is unwanted or damaging to caregivers. Therefore it has to be repressed or dissociated or otherwise defended against in order to protect the ties with caregivers. So it led to a reformulation of conflict along those lines. It also led to a reformulation of the so-called dynamic unconscious. The dynamic unconscious within this framework pertained to those affect states that could not be allowed full articulation because they were inimical to a needed tie with a caregiver.

So instead of being an unconscious compartment or a subterranean zoo—you're supposed to chuckle at that one—in which forbidden drive derivatives were held captive, the concept of dynamic unconsciousness referred to affect states that were not to be allowed to come into full being because they would jeopardize a needed tie.

Emotional trauma, as I alluded, pertains to what affect states feel like when they are not met with a needed attunement or responsiveness, or what I later came to call a *relational home*. They are experienced as overwhelming, disorganizing, disruptive, dangerous. This is a way of understanding trauma in terms of the vicissitudes of affect within an intersubjective context. In general we might say that the overall significance of this article, which was very significant in the development of our framework, was to show that emotional experience is inseparable from the intersubjective contexts of attunement and malattunement in which it is felt. Emotional experience is inseparable from its context.

The effect of this understanding is to contextualize everything. That is, the move from instinctual drive to affect as the motivational prime mover contextualizes everything. Because affect states, unlike drives, which are components of a Cartesian isolated mind, are something that from birth onward are integrated or not integrated within relational contexts. So if you put affect at the motivational center you end up contextualizing everything; both the phenomenological and the contextual components of phenomenological contextualism got a big boost from that article, "Affects and Selfobjects."

PM: Would you say that once you wrote "Affects and Selfobjects" in 1985, the major building blocks of your phenomenological contextualism were kind of put into place for further refinement and elaboration? Does that sound right?

RS: I think so, yeah, right. I think you can see that in terms of specific chapters and articles. First was the last chapter in *Faces in a Cloud*. Then there was the article on psychic structure. Then there was the joint article with George on the phenomenology of the therapeutic situation. And then there was the article on borderline phenomena by Brandchaft and me. And then these two last articles, the one on transference as organizing activity with Lachmann and "Affects and Selfobjects" with Dede.

Right. I think the basic elements of the framework were in place with the publication of those two articles in 1985, especially "Affects and Selfobjects."

Further development of the framework took place as a result of my immersion in phenomenological philosophy, which became more intense as the years passed. Coinciding with this immersion was my own experience of a terrible trauma—the death of Dede in 1991. There was a conference in 1992 in which I relived the trauma of finding her dead.

George's and my newly published book, *Contexts of Being,* was released, and a batch of initial copies was delivered to the conference where I was a panelist. I picked up a copy of the book and whirled around to show it to Dede, because she would be so happy to see it. Of course she was nowhere to be found, because she had died 20 months before. That experience of whirling around to show her the book and finding her gone took me right back to the experience of waking up one morning and finding her dead, what I later came to call a *portkey to trauma.*

The features of my state when that happened were a sense of isolation and estrangement from everybody at the conference, a sense of meaninglessness, a sense of being alien and different from everybody else. I struggled over the course of several years trying to figure out what that state was about. And some of the readings I was doing were helping me with that. Reading Gadamer, for example, on trying to understand a world that is incommensurable with one's own.

PM: So you were reading Gadamer and perhaps other philosophers in the 90's?

RS: Yeah. And I felt that I was from a different world than the other people at this conference. That they could not possibly understand my experience because we were from different worlds, different worlds of experience. What I came to understand was that this state that I was in was a product of having what I call the *absolutisms of everyday life* shattered. All the illusory presuppositions that give one the feeling that one is safe in the world are shattered by trauma. I concluded that it lay in the essence of trauma that it entailed a shattering of these reassuring illusory assumptions about oneself and the world. A shattering of all sense of safety and continuity.

PM: That sounds like it would place the experiences of anxiety and grief right at the center of much trauma.

RS: Oh yes, definitely. I wrote that in an article in 1998 and it was published in 1999. In 2000 I started a leaderless philosophical reading group that ran for two years. The first year was devoted to a careful reading of Heidegger's (1927/1967) *Being and Time*. When I came upon the sections in *Being and Time* in which Heidegger described his existential interpretation of anxiety, I nearly fell off my chair! The phenomenology that Heidegger described was almost identical to the phenomenology of the traumatized state that I had experienced at the conference in 1992 and that I had written about two years earlier.

I quickly realized that Heidegger's existential philosophy as presented in *Being and Time* provided invaluable philosophical tools for our psychoanalytic phenomenological contextualism. For one thing, Heidegger radically contextualized the Cartesian isolated mind. Descartes conceived of the mind as a self-contained thinking thing, ontologically separated from its context, its world. Heidegger argued very persuasively that our existence is always a being-in-the-world, in which our experience of ourselves is inseparable from our experience of the world, our context, our situation.

Heidegger's ontological contextualism provided a perfect philosophical grounding for our psychoanalytic contextualism with which we could replace the Cartesian isolated mind. But perhaps even more important for me at the time, Heidegger's analysis of anxiety and his discussion of world-collapse, uncanniness, and authentic being-toward-death provided invaluable philosophical tools for understanding the existential significance of emotional trauma and for working with traumatized states.

So I decided at that point to go back to graduate school and work on a doctorate in philosophy. I wrote a number of articles, two books, and a doctoral dissertation on Heidegger and what George Atwood and I came to call *post-Cartesian psychoanalysis*—psychoanalysis with the Cartesian isolated mind expunged.

So at that point I married my mistress, philosophy, and became a philosopher, although I was never able to get an academic job as one. I began to think about a number of philosophical issues and implications of the developing framework of phenomenological contextualism.

For instance, I began to think, with George, about how to understand metaphysics and metaphysical systems. We came to recognize that psychoanalytic metapsychologies were actually forms of metaphysics, because they all formulated absolute realities and universals that were presumed to be true of everyone.

Our critique of these metapsychologies, or we might say our critique of these forms of metaphysics, understood them as forms of illusion. We were influenced by the work of Wilhelm Dilthey on this issue, who saw that these forms of illusion represented evasions of human finitude. Human limitedness and transience, context-dependence, existential vulnerability—all of these were evaded through the process of absolutizing and universalizing various ideas and transforming them into eternal everlasting entities.

PM: So metaphysical illusions, including metapsychologies, are attempts to answer human contingency with absolutes and thereby counteract those affects, like the complexity anxiety we discussed earlier, that disclose the limitedness of the human person? What other regions of emotional life do they serve to counteract?

RS: Well, I think they serve to counteract every region of emotional life, because emotional life is itself always context-dependent. It is contingent and vulnerable. I don't know that I would want to specify a region. I think it's more like an evasion of these characteristics of emotional life as such.

PM: Right, so what's counteracted is inherent to all emotional life, namely, the "unbearable lightness" that runs through all subjective emotional experience?

RS: Right. Or as we put it in the book, *Contexts of Being, the unbearable embeddedness of being*. Thanks for reminding me of that.

PM: But, just to clarify, you're supposing that this embeddedness structures all experiencing, even if prereflectively, and can either come into experience voluntarily or per force of trauma, or be counteracted and evaded? Am I correct that you think of the embeddedness of human being and experiencing as disclosed in emotional experience, whether or not we own it?

RS: Yes, exactly, because I think emotional experience constitutes the heart of being itself.

Okay, so coming back to your original question about why I don't identify as a psychologist, that should be clear, right? In terms of the traditional accepted model of the clinical psychologist as a scientist-practitioner.

Instead I want to see our framework as a form of applied philosophy, applied phenomenology, or phenomenological inquiry, aimed at the particular rather than the universal. I think one of the side benefits of this perspective is that it encourages a critical attitude toward traditional psychology.

PM: Can you formulate that attitude?

RS: Well, you know, one of the slogans that prevails in the field of clinical psychology is *evidence-based treatment*. Never mind that this is something that is strongly influenced by the insurance industry. This is the banner that is waved over and over again in traditional psychology, evidence-based treatment, which means treatment methods that are based on traditional scientific research. What I described before as identifying variables, quantifying them, measuring them, doing statistical analyses, and so on.

This is the nature of the evidence that people are talking about, and there's little or no philosophical questioning of what the proper evidence is to guide the therapeutic approach to a suffering human soul. Is this the kind of evidence that's meaningful in helping us to come up with a therapeutic approach to human suffering? Or is there another kind of evidence that is much more germane, the kind of evidence that takes me back to my roots in personology—evidence that comes from the careful investigation of the experiential world of the individual and the contexts in which it has developed.

So the perspective that we've been developing I believe has very important implications for reflecting on the kind of evidence that we should be looking for and studying as a basis for so-called evidence-based treatment. I was going to say, how's that for a note to end on?

PM: Well, that sounds great. Two questions I want to ask before we end are these: what direction do you envision for your future thought, and what do you deem as valuable areas for future research for those people, like myself, who have been influenced by your intersubjective-systems perspective? I suppose you've already identified one such research area, namely, phenomenological-contextualist inquiry into the forms of "evidence" that help promote and evaluate clinical approaches to emotional suffering.

RS: Yeah, I think there's a lot of work to be done in that area. And it also happens to dovetail with a theme in Heidegger's later work, which in effect was a critique of what he called the *technological way of being*, of which scientism is an example. That is, the assumption that the methodology of the natural sciences, which I described before, you know identifying variables, and so on, is the only legitimate path to knowledge and truth. That's the assumption of scientism.

PM: Right.

RS: It's an assumption that deserves a lot of philosophical reflection and critique.

PM: Again, whether with respect to your own future work or that of those who might extend your intersubjective-systems perspective, any further thoughts on what area of inquiry might have the most interest to you as a focus?

RS: Well, here's something that George and I have been discussing—a radical critique and deconstruction of the entire system of psychiatric diagnoses, which is basically a system for diagnosing sick Cartesian isolated minds. That whole framework needs to be deconstructed and replaced, we're thinking, by a phenomenological description of differing experiential worlds and their contexts of origin.

And what I would expect would come out of a study like that is that it would replace diagnostic entities with traumatizing contexts.

PM: I'm remembering something you said at a conference that really impacted me. You said, as I recall, "There are no diagnostic entities, only traumatizing contexts. "Am I quoting you accurately?

RS: You are.

PM: Let me loop back to your bigamy and add something before we close. Am I evading the finitude of our conversation? Holding in mind that you have identified psychoanalysis as a form of "applied philosophy," it can also be said that philosophers necessarily organize their philosophical systems with their own particular life experiences. So the domain of psychoanalysis is always already in philosophy. Maybe philosophy and psychoanalysis are a unitary discipline in some sense, even if philosophical and psychoanalytic activities occur out of different levels of that unity?

RS: Yeah, I think that's right. Well, it's similar to the distinction between philosophical phenomenology and psychoanalytic phenomenology. The philosophical being aimed at the general and the psychoanalytic being aimed at the particular.

PM: Right. Just to reiterate, it seems to me that when you look closely at the general you can't really talk about it without reflecting on one's own personhood, and vice versa.

RS: Right, absolutely. So the philosophies of the general need to be contextualized by an investigation of a particular.

PM: Right. That seems to me like it must be the case.
RS: The experiential world of the philosopher and its context.
PM: Among the things that have really influenced me in your work has been your recent demonstration of the mutually enriching relationship between philosophy and psychoanalysis.
RS: Right, definitely.
PM: So thanks very much.
RS: Speaks in favor of bigamy, right?
PM: Absolutely. Except I find the notion of disciplinary "bigamy" less favorable than something that sees these domains of inquiry as unified, as not really so divided. I'd like to find a way to capture such an idea in our language. I think that might be valuable.
RS: Good point. I'll pass the baton to you.
PM: Thank you for your time.
RS: You're welcome, it was a pleasure.

Bibliography

Atwood, G. E., & Stolorow, R. D. (1984). *Structures of subjectivity: Explorations in psychoanalytic phenomenology*. Hillsdale, NJ: The Analytic Press.

Atwood, G. E., & Stolorow, R. D. (2014). *Structures of subjectivity: Explorations in psychoanalytic phenomenology and contextualism* (2nd ed.). New York, NY: Routledge.

Atwood, G. E., & Stolorow, R. D. (2016). The phenomenological circle and the unity of life and thought. *Psychoanalytic Review, 103*, 291–316.

Atwood, G. E., Stolorow, R. D., & Orange, D. M. (2011). The madness and genius of post-Cartesian philosophy: A distant mirror. *Psychoanalytic Review, 98*, 263–285.

Heidegger, M. (1967). *Being and time* (J. Stambaugh, Trans.). Albany: State University of New York Press. (Original work published 1927).

Klein, G. S. (1976). *Psychoanalytic theory: An exploration of essentials*. Madison, CT: International Universities Press Inc.

Murray, H. A. (2007). *Explorations in personality*. New York, NY: Oxford University Press (Original work published 1938).

Orange, D. M., Atwood, G. E., & Stolorow, R. D. (1997). *Working intersubjectively: Contextualism in psychoanalytic practice*. Hillsdale, NJ: The Analytic Press.

Socarides, D. D., & Stolorow, R. D. (1984/1985). Affects and selfobjects. *Annual of Psychoanalysis, 12/13*, 105–119.

Stolorow, R. D., Atwood, G. E., & Orange, D. M. (2002). *Worlds of experience: Interweaving philosophical and clinical dimensions in psychoanalysis*. New York, NY: Basic Books.

Stolorow, R. D. (2002). From drive to affectivity: Contextualizing psychological life. *Psychoanalytic Inquiry, 22*, 678–685.

Stolorow, R. D. (2007). *Trauma and human existence: Autobiographical, psychoanalytic and philosophical reflections*. New York, NY: Routledge.

Stolorow, R. D. (2011). *World, affectivity, trauma: Heidegger and post-Cartesian psychoanalysis*. New York, NY: Routledge.

Stolorow, R. D., & Atwood, G. E. (1979). *Faces in a cloud: Subjectivity in personality theory*. Northvale, NJ: Jason Aronson.

Stolorow, R. D., & Atwood, G. E. (1992). *Contexts of being: The intersubjective foundations of psychological life*. Hillsdale, NJ: Analytic Press.

Stolorow, R. D. & Lachmann, F. M. (1980). *Psychoanalysis of developmental arrests: Theory and treatment*. Madison, CT: International Universities Press.

Stolorow, R. D., Brandchaft, B., & Atwood, G. E. (1987). *Psychoanalytic treatment: An intersubjective approach*. Hillsdale, NJ: The Analytic Press.

Peter N. Maduro, J.D., Psy.D., Psy.D. is a clinical and forensic psychologist, and psychoanalyst, in private practice in Santa Monica and South Pasadena, California. He is on faculty and a Supervising and Training Analyst at the Institute of Contemporary Psychoanalysis in West Los Angeles, California.

Thinking Psychology Otherwise: A Conversation with Mark Freeman (Interviewed by David Goodman, Boston College)

Dr. Freeman is a remarkable and expansive thinker, teacher, administrator, editor, and scholar whose voice and manner of speaking have facilitated conversations that are rare and precious. In each of his roles, Dr. Freeman calls for greater "fidelity to human experience" (to use his words) and inspires his colleagues, students, and readers to attune more closely to the wild expanses of life. As a narrative psychologist, Dr. Freeman continuously defies disciplinary boundaries and creates possibilities for "thinking otherwise" about our identity and way of being in relationship to ourselves and others.

Dr. Freeman's leadership roles span from his position as Chair of the Department of Psychology at the College of the Holy Cross to his work as Editor of the *Explorations in Narrative Psychology* book series at Oxford University Press to his recent presidency of Division 24 (Society for Theoretical and Philosophical Psychology) of the

M. Freeman (✉)
College of the Holy cross, Worcester, MA, USA
e-mail: mfreeman@holycross.edu

H. Macdonald et al. (eds.), *Dialogues at the Edge of American Psychological Discourse*, Palgrave Studies in the Theory and History of Psychology, DOI 10.1057/978-1-137-59096-1_6

American Psychological Association. In recognition of his work and far reaching contributions, Dr. Freeman was the winner of the 2010 Theodore R. Sarbin Award and most recently (in 2016) the Steve Harrist Distinguished Service Award. He is a fellow at the APA and has been an integral figure, along with Kenneth Gergen and Ruthellen Josselson, in bringing qualitative methods into mainstream conversations and even into greater institutional and scholarly recognition within the APA.

Dr. Freeman currently serves as Distinguished Professor of Ethics and Society at College of the Holy Cross, where he has been on faculty for several decades. Dr. Freeman's students enjoy his highly interdisciplinary courses where they read everything from Sigmund Freud to Ta-Nehisi Coates to Simone Weil. His students describe his style as simply life-altering and an invitation to greater truth and beauty.

In addition to his editorial work at Oxford University Press and on several journal boards, Dr. Freeman's own scholarly achievements are extensive and his writing style is widely recognized for its movement between personal narrative, complex theory, and evocative prose. He brings together thinkers who are seldom explored in relationship to one another and who even more rarely are put in conversation with the psychological discipline. He has published over ninety book chapters, articles, and monographs. His books include *Rewriting the Self: History, Memory and Narrative* (1993), *Hindsight: The Promise and Peril of Looking Backward* (2010), and most recently, *The Priority of the Other: Thinking and Living Beyond the Self* (2014c).

DG (David Goodman): It is no small thing to me to be the one sitting here both introducing Mark Freeman and diving into some of the critical questions that his work has brought to us and to this conference in particular. Mark, actually, was the very first plenary speaker at the Psychology and the Other Conference in 2011, inaugurating this entire event.

In the last five years, Mark has become a generous mentor to me, and our friendship has become something most precious. I have a deep and affectionate respect and reverence for Mark's work and his presence in the field of psychology. His path and approach have been an inspiration and it's an honor to sit with him today.

I had the great pleasure of recently co-editing an Oxford book with Mark titled *Psychology and the Other*. I must admit that writing with Mark is like training with an Olympic athlete. It's quite extraordinary to watch someone that has the conceptual range and mastery of language that he does, all the while maintaining a hospitality toward authors and writers that they do not always deserve. In the editorial process, we would have frequent conversations about how to live out a Levinasian ethics while communicating a scathing critique! I learned a lot from Mark in that process.

One of the most compelling aspects of Mark's work is the way that he straddles conversations happening in narrative theory, the humanities, continental thought, research methodology, aesthetics, psychoanalysis and ethical phenomenology. If you quickly review his speaking docket and publication titles in any given year, you'll be struck by his tremendous range, all the while remaining one of the deepest and most precise thinkers and writers that I know. Mark has a genius capacity to metabolize and translate disparate worlds and then write about them with a literary and personal voice that illuminates it at the level of life.

With no further ado, let's spend some time in questions that bring us to the central concerns of this interview: how Mark understands the field of psychology in its current form, his beliefs about where the field could be going, and his hopes with regard to these future directions.

But before we get to all that, I thought we could start a little bit more personal. Of all the fields you could have chosen, why did you tether up to the modern discipline of psychology? And, what was your road into the psychological world?

MF (Mark Freeman): I'll address that in just a moment. First I want to thank you for that kind, gracious and somewhat embarrassing introduction. I also just want to say what a great pleasure and privilege it's been getting to know you and working with you. We've had occasion to go away on fun weekends with our mates—working hard during the day and partying hard by night—all that's been great.

And I just want to congratulate you on the amazing work that you've done for this conference, for all three of them. And this one, again, is just an extraordinary event. One word that comes to mind—there's a lot of words that come to mind about David—is gracious. In fact, I have to confess to you, yesterday at one of the plenaries, at the end of it, somebody announced that there would be cookies, right? And I immediately had an image of David in a hot kitchen with an apron on baking the damn things, honest to God. I don't want to know if that's true.

DG: I'll let you have your fantasies!

MF: So how did it all begin? You know, I knew you were going to ask me that question. I must tell you I find that to be a truly vexing question. How is it that we come to choose our professions or why it is we come to be this kind of psychologist rather than that? Those, I think, are mysteries, in a way. But since you asked me, I'll venture a couple of things.

I can talk about fascination with human behavior and the inner workings of things. Like many people in adolescence, you're fascinated with what makes people tick and not tick and all the rest. That was certainly a part of me. But you know what? That was true of all of the people I hung out with at the time and they went on to do quite different things.

Truthfully, I think pain was probably a part of it. There are a number of different things that happened, especially during the high school years, that undoubtedly left an imprint, although what kind of imprint is difficult to say. When I was 17, I was almost killed in a car accident that left me in a hospital for a month with lots of broken bones and a new sense of finitude and mortality and so on. It wasn't too long after that that a good friend of mine died at age 17. And not too long after that was my father. It's not like I was in a pit of despair, but I do think that those things have a way of opening you up and exposing regions of experience, regions of feeling, that somehow need attention.

So, that was certainly part of it as well. There were always questions about whether psychology was the right discipline, and if I had to identify the two other areas of thought that I would have been most attracted to, I'd cite philosophy and literature. Those still remain passions. Students, given my approach, will often ask, "Why psychology given your obvious love of philosophy and literature and art and so on?" And my answer is that actually psychology, at its best (and I realize there are a lot of variants of best), is an extraordinary undertaking. To really explore the human condition in its variability, its depth, its richness, its ugliness, its beauty and more, is an amazing opportunity.

You can do that to some extent in philosophy, obviously, but I found myself attracted not only to thinking about some of the basic philosophical categories—consciousness, man, being, and so on; I was attracted to thinking about all of those issues through *this* person or *that* person, this concrete flesh and blood being who instantiated all of those different categories. I loved literature as well, and continue to. But there was also something about that pursuit that didn't quite get me where I needed to go.

So, I landed in psychology. It was a shock to the system. It was absolutely not what I imagined it to be. I was certainly tempted to flee, and my guess is that that was the case with a number of people, and probably still is. It was kind of interesting in some ways, in the same way that tinkering with machines can be interesting. It was the heyday of behaviorism and I didn't know what to make of it all, I really didn't. But I was fortunate enough to stumble onto a couple of courses in philosophy that allowed me a different and a much more compelling image of what psychology might be.

DG: Can you speak to that? What courses, ideas, professors, and experiences were major influences on your development during your time at the University of Chicago? If you wouldn't mind, walk us into this a bit. What shaped you?

MF: In college, there were several courses but my main entry into this basic mode of thinking was a course in "Phenomenological Psychology," which I took with a Merleau-Ponty scholar,

> Martin Dillon, in the Philosophy department. That was amazing. It was just like "Wow, that's what I thought psychology was!" And I took another course called "Visual Thinking," which looked at the work of people like Rudolf Arnheim and got me thinking about issues of perception and reality and where art fit into all of that. So, that opened up a world. It was not easy to figure out where I would go, or for that matter *could* go from there because as we all know, there aren't that many programs in contemporary American psychology that would allow one to pursue those kinds of issues.

It took several years to try to figure out what was what and have a good time while I was doing it. I eventually landed at the University of Chicago, partly because of what was going on in the department, or the "committee," that I eventually became a part of. I wound up earning my Ph.D. in the Committee on Human Development, where people would study the various social sciences along with whatever else they wanted to pursue. We had to take some 27 courses, and only 9 of them were required. So, you could kind of be whatever you wanted—at your peril. I say "at your peril" because if you wound up being too heterogeneous and then you looked for a job in some mainstream department, it may not fly. But it was an amazing stroke of luck in many ways. The Committee on Human Development was devoted to studying the course of lives in all of their complexity and multidimensionality. That meant we had to learn something about not only psychology but also biology and sociology and anthropology and so on. I also very much wanted to continue my work in philosophy.

So, I applied to study with Paul Ricoeur (you had to apply to some of his courses to get in). I was the lone social scientist who did that, and I was able to get into a course early on, a two-semester seminar in the "Phenomenology of Time Consciousness," followed not too long after that by a co-taught course, featuring Ricoeur, Steven Toulmin, David Tracy and Langdon Gilkey called "Historicity, History and Narrative." I was able to take another course with Ricoeur and Mircea Eliade called "Mythical Time." *Explosive.*

So it was the right time and the right place and those ideas are still very much with me. As you know, I wound up doing a lot of work in narrative and time and all that kind of stuff (e.g., Freeman 1984, 1985a, 1985b; Freeman et al. 1986; Freeman and Larson 1986). It was a heady time to be there, and I look back on that epoch with immense gratitude for my teachers, for my friends, for my wife, for Chicago and more. It was an unbelievable thing, it really was.

Now, the question, of course, after that, is where do you go from there? Some people from Chicago who were topnotch scholars fell between the cracks of the disciplines and weren't able to get positions, which was really unfortunate. But in 1986, in the APA (American Psychological Association) *Monitor*, I actually saw an ad for a phenomenological psychologist (which I don't consider myself in any strict sense of the term) to teach "History and Systems of Psychology" and whatever the hell else I wanted to do.

DG: You don't see those ads anymore.

MF: You see those ads once every decade or something like that. I'm immensely grateful for that, too—to be able to have landed at Holy Cross and do what I do.

DG: On more than one occasion, I've actually heard you express your appreciation for Holy Cross and the incredible students and colleagues that you have worked with over the years. I've also heard you say that many of your colleagues, particularly in the Psychology Department, don't always know what to make of your work and how you fit exactly. You're allowed to do your thing, but it's not necessarily understood as a part of psychology proper. Part of why I look forward to hearing your response is because I imagine your answer will illuminate some of the larger questions regarding where the field of psychology is right now. So, why don't they know what to make of you? Even though they love you and you're their chair!?

MF: Yeah, I'm the chair; it doesn't matter what they make of me! It's a benign relationship and some of that benignness is probably borne out of our mutual ignorance. You know, I jokingly have said before that one of the reasons they gave me tenure and

> promoted me is that they probably didn't read the stuff. I don't actually know if that's true, so we should probably scratch that from the record! But, you know, it's a quite traditional department. Because Holy Cross is a Jesuit school that prizes religious and philosophical questions, there was a space for the likes of me. But, what I do is largely alien to them. They approach it with some measure of fascination sometimes, but they still don't know quite what to do with it.

The story of my promotion report is illustrative. I think it was 1998, I'm supposed to receive my promotion report and a guy in my department comes up to me and gives me the report. There's a letter on the top and it says something like, "Dear Mark: As always, we enjoyed reading your work. But we have absolutely no idea how to gauge its significance for psychology. Consequently, we're passing it on to the Department of English because they may know better what to do with it than we would." My immediate response: "Come on! God, this is so bizarre!" They were just kidding. But there was a truth there. I've sometimes been asked, why do you call yourself a psychologist? And I used to have pretty ready answers for it. "I'm a scientist, too. I'm interested in exploring reality. I just do it in a different way than you." There's still some sense in which I'd be prepared to say that. But, I suppose I call myself a psychologist because I'm in a department of psychology. Fortunately, I have the liberty at this point not to put myself in any particular container, but instead to just do the work that interests me and that draws me forward.

It's strange, and it's been a challenge to navigate this sort of thing. It's a challenge to navigate my department. It's been a challenge to navigate the discipline of psychology more generally. Qualitative work is just one example. When we (Ken Gergen, Ruthellen Josselson, and I) were initially considering joining a division of the APA (we had been asked to consider joining the Division of Evaluation, Measurement, and Statistics!), there were loud protests at different points. You know, "This is anti-scientific! It's the end of rigor as we know it!" All we wanted to do was basically say, "You know, we're interested in studying people and we want to do that in lots of different ways as a vehicle for doing better justice to human reality and the human condition." But that took a lot of argumentation. The idea that we would seem renegades in the name of exploring the human condition is, of course, profoundly ironic. But, that's much of the way it was.

DG: You speak a lot in your work about the desire to maintain fidelity to experience. This is something that one would imagine the field of psychology would hold as a primary concern. Regrettably, for a variety of reasons, this commitment does not appear to receive the consideration and attention that we would hope. What are the factors that you believe contribute to this state of affairs? What are the current impediments? What is it about the condition of our field that makes this a difficulty?

MF: This is probably too general a gloss, but for the most part, psychology, in its aspirations to scientificity, has used methods and methodological approaches that try to *encapsulate* phenomena, by using questionnaires and one-way mirrors and experimentation and all the rest of it. The idea—and this has been around since the beginning of psychology—was that experience itself was somehow too recalcitrant, too ambiguous and messy, and that those dimensions of the human condition would be better left to poets and philosophers and others better prepared to do that sort of plumbing, so to speak. The way I had framed it earlier, in fact, is that partly because of psychology's scientism, it actually wound up being *less* scientific rather than more because much of it failed to be adequate to the object of inquiry—namely *us* (e.g., Freeman 2007a, 2007b, 2011). So there's a certain sense in which I see qualitative work and the work that many people in this room do—this will sound kind of audacious—as a vehicle for psychology's *self-realization* (Freeman 2014d, 2015). It seems to me that it's still a somewhat immature science, an immature undertaking. Now, I'm not sure it all has to be formulated as science; that's another issue. But its fetish for a certain kind of objectivity has led it to want to contain things and measure them and quantify them.

I have no particular problem with this when it's appropriate. My aim has never been to replace that whole structure, but to augment it, to supplement it. As I say to my students, the problem so much isn't what's included in psychology; it's what's been excluded, often vehemently. Those are the impediments. And they're still there. I like to think that we're in the process of taking some baby steps toward crafting a discipline that's more appropriate for some of our aspirations.

DG: Taking another angle on that same question: you've called for a more "capacious science" and sometimes even a "poetic science." I believe you just described that well. What is threatening about that to the psychological science? Why is it that giving voice to this in the context of an organization like the APA is often experienced as though you are undermining an important edifice?

MF: Once you start using words like "poetic," in much of psychology it probably comes off as sort of soft and touchy-feely. I tried to circumvent that immediate connotation by tacking on the word "science" (e.g., Freeman, 2011, 2014d). It's come to feel a little bit disingenuous (see Freeman 2015). But let me say just a word about what I meant by poetic science. Not only are many of the methods of psychology objectifying and problematically encapsulating, but so is the language. I mean, most writing in psychology is kind of dead—partly because of its aim of keeping a distance, and partly because of its insistence on using a kind of precision that its aspirations to science warrant. But the fact is, oftentimes more literary and more poetic language is more adequate to experience than that kind of language (e.g., Freeman 1999, 2000b). So, once again, there's a paradox here: through becoming more *artful* with how it is that we go about doing our work and representing it and speaking about it, we might actually wind up being more *scientific* in a way.

There's a passage from (Breuer and Freud's) *Studies on Hysteria* (1895/1955) that I often read to my students where Freud confesses a kind of embarrassment, almost, about what psychoanalysis—and psychoanalysts—seem to have to be. He says at one point (and I'm paraphrasing), "I began my career as a neurophysiologist. I was involved in electro-prognosis, all this sort of stuff, but it didn't get me anywhere. And strangely enough, when I started to make use of the kind of strategies that writers use, I was able to gain some kind of illumination into the phenomena." The only "consolation," he went on to say "is that the subject matter necessitates it." And so, as you know, psychoanalysis became a kind of narrative undertaking, where in a sense people become seen as works of literature (e.g., Freeman 1985b, 1989, 2002, 2007b). It's partly because that seems to be what we are. I realize there are lots of different ways of framing human experience, so I don't want to over-elevate that. But Freud became aware that it might be necessary to use an entirely different kind of writing and a different way of thinking about the human condition in order to be adequate to it. And that maybe that's what *science* had to be (Freeman 2007a, 2007b).

DG: That's very interesting and powerful. I would like to take us in a different direction, to some of your more recent work, particularly your book *The Priority of the Other* (2014c). In a rich, balanced, and deeply dimensional manner, you call for the displacement of the cogito and the prioritization of the Other. I know this is an enormous question, but what are the cultural, historical, political, and theoretical concerns that make it difficult for American psychology to be receptive to a prioritization of the other? You have a vision for an *ex-centric* psychology, one that allows for a self that reaches outside of itself and prioritizes the Other. First of all, why? Second, what stands in our way of that? Perhaps, you can first flesh out what you are even hoping for.

MF: Again it's a bit audacious, but what I'm hoping for or aiming for in that book is rethinking some of the most basic categories through which we understand psychological experience. I think many of the categories that we use are actually belied by experience. There are lots of reasons that we could point to (and people have written about them, many in this very room)—about the perniciousness of certain strands of individualism or the culture of narcissism, all of that sort of stuff. But it seems to me that we

> also have a kind of culturally constituted self-understanding of who and what we are and that even if we begin to adopt more relational language or dialogical language or what have you, it isn't quite adequate to what it is that we seem to be (Freeman 2012, 2014c). Even here, I can't divorce all the theoretical musings that we all love so well from the experiences that I have in the world. The experience of my wife being diagnosed with a disease for instance—she's doing well, as you know, and is a vital person—having that reorient me in an immediate and visceral way and allowing me to see what's most significant.

One of the things I suggest in the book is that the default mode of being for much of our lives, given the way we lead our lives, is what I call "ordinary oblivion." We're caught up in ceaseless movement of everyday life and oftentimes it takes an illness of a loved one or a death or who knows what to be able to shake us awake. I'm not interested in talking about altruism. That's not what the book is after. What I do suggest is that our lives seem to be characterized by competing forces, one of which moves inward toward the ego, centripetally, and another that moves outward or is drawn outward, toward the world—by others, by nature, by art, by God, and so on. So, it's not only the importance of other people that I want to underscore. I do, as you know, make significant contact with people like Emmanuel Levinas and Martin Buber, thinkers who are particularly prominent at this conference. But I also make significant contact with thinkers like Iris Murdoch and Simone Weil, who emphasized, for instance, the way in which a work of art can somehow serve to arrest some of our own egocentric preoccupations and alert us to what's there in the world.

There is another set of issues that I've been interested in for a long time and that I know other people in this audience share. And it has to do with the kind of otherness, so to speak, that many of us encounter in the context of aesthetic experience, or the experience of music, or the experience of a new grandchild or the experience of love, or what have you. What *is* that? (e.g., Freeman 2004, 2009a, 2014c, 2014b). There are a number of ways that we can encapsulate those kinds of experiences, too. We can do it neurobiologically, we can do it psychoanalytically, and so on. But another inspiration for the book is William James and other people who want to say it may be that some of the otherness that we encounter in those experiences is *not* just a product of the psyche, is not a projection, but in some sense exists in the world. It may be that a fully naturalistic account of the self that doesn't recognize our connection to those forms of otherness is a kind of shrunken image of the self and not entirely adequate.

That led me to think about words like "ex-centricity" rather than egocentricity, the notion being that maybe the self isn't quite as bounded as we've imagined it to be. Again, I know that others in this room have spoken about that. But I'm very interested in the kind of energies beyond the perimeter of the self that draw us beyond ourselves.

A lot of this is exploratory. It's not as if I have a fix on these kind of things. In many ways, what I try to do in this book is ask, Can this way of thinking and speaking allow us to see some phenomena in a new and maybe more adequate way than we had been able to see them before? We're not used to that way of thinking, certainly not in psychology. Among the historical factors we could talk about__autonomous individualism and neoliberalism and all the rest of that__it's pretty familiar territory. I think we could also look to at least certain strands of Darwinism, if not Darwin—the notion that we're self-seeking beings whose primary interests are ultimately egocentric and that even moral life may well be a function of the investments we make in others because of the return we hope to get, whether that's conscious or not. I'm not convinced that all of that is valid despite the fact that much of it has become axiomatic. I think it's worth thinking against those things.

But I don't only want to think against. Some of you will remember from the midwinter meeting that I wanted to call a symposium "Critique and Creation" just for this reason. In Division 24, which is where many of us work, we're very good at critique. That critique is vitally important—uncovering assumptions and all the rest of it. But I'm equally interested at this point in trying to imagine what kind of psychology we could build.

DG: Yes, indeed! One of the points that you make toward the end of *Priority of the Other* (which I really enjoyed) was a reference to the "service to the Other" and social justice oriented mentalities that can often miss the point in contemporary culture. When you speak about the ways that service can quickly become thinned out and stale, I think about David Brooks's (2016) recent book, *The Road to Character*, where he starts with the question of whether we want to live out of a eulogy ethics or resumé ethics. A great deal of service work, especially for undergrad students who are all competing with each other, is oriented toward resume virtues.

But, back to the point in your book: you don't leave the reader with a simple critique of this phenomenon. You argue that attention and devotion need to be cultivated and nourished in our culture to facilitate the proper means by which we can get out of and beyond ourselves and hear the Other. You are not critiquing service work, but you are concerned about whether they are driven by the wrong engines. And you actually have the sense that the lack of attention and devotion in our culture is part of what contributes to this.

MF: One quick comment on the ethos that you mentioned—Holy Cross and BC, where the motto is "men and women for and with others," which is lovely. Another reason why I've come to be interested in these ideas because of the people who are there, like my friend Jim Kee from the English Department and John (Manoussakis) from Philosophy and others who take all this stuff very seriously and have allowed me to see a validity and a significance in the religious that, quite honestly, I really hadn't seen before.

But that motto can sometimes become too much of a container, with the result that a student at Holy Cross might think that unless they were doing work in a soup kitchen or going to Guatemala for spring break, they're somehow not valid enough. One of the things I try to suggest to them is that being a painter or a poet, whatever, also is about a kind of service to the other, so let's not think of what it means to be for the other as just involving an explicit crusade for justice. It has to do with being attentive to the world and speaking to it in some meaningful way, whether that way is social action, whether that way is creating art, whether that way is becoming a particular kind of psychologist, a philosopher, or what have you. But how to put all this into practice? That's not easy. The chapter of the book that I put off for a number of years, actually, was the final chapter. What does all this add up to? The last thing I wanted to do is put out truisms about how we might become more other-directed or something. I mean, I'm interested in that, but there are no simple strategies for moving in the kind of direction that I'm suggesting here.

DG: That's helpful. You're not going to like this next question! If you were to be assigned, or sentenced, to be president of the APA next year, what are several initiatives that you would take on with the hopes of contributing to "righting" the direction of the discipline?

MF: Much of contemporary psychology remains a regime and it's a regime into which students are socialized, and in a way that's very, very problematic. Again, there are all kinds of reasons for that. It's a young science, it has scientific aspirations, whatever. But the fact that it still remains a discipline that's not entirely fit for human habitation, and that many students see that and feel that, is a problem. I mean, look, one of the small goals that we were looking toward as we formed [the Society for Qualitative Inquiry in Psychology] was the possibility that a research methods course might have a section in it that was more attentive to these kinds of issues. Contemporary psychology is bound up with all kinds of commercial interests and industrial interests and so forth that are problematic in lots of different ways that I wouldn't care to enumerate here. I also think the discipline suffers from a lack overall of (this will sound crude) intellectual integrity. I don't mean to suggest that there aren't people, lots of them, in the discipline, who have that. But there's a whole lot of stuff that goes down in the discipline that basically has to do with churning out work that goes into "the best journals," and building the kind

> of CV that's going to get you a reputation, that's going to be marketable, and so on. That's true in all disciplines. But I must say, I think there's an extra crudeness in much of psychology. So that's part of it.

I also think that we would do well to continue to rethink some of the most foundational assumptions that the discipline operates with. Methodological, for sure, such as the problem of interpretation, how to have that be more focal to the discipline. But I also mentioned the whole idea of a naturalistic framework and the almost axiomatic assumption at this point that we are fully material beings and that neuroscience is going to pave the way to the future by being able to illuminate that.

I think there are many people who on some level are in the process of making psychology, as we have known it, obsolete. There continue to be books coming out like *The Illusion of Conscious Will* (Wegner, 2003), *The Self Illusion* (Hood, 2013), and so on. These illusions have supposedly been *revealed*, discovered, by the latest findings in neuroscience and other such pursuits. I have no interest in dissing neuroscience. There's interesting work, and it's vitally important. But psychology is always in the business of establishing a new kind of way of thinking and having that colonize everything. There's a lot of room in this discipline, given the varieties of the human condition as well as the plural methodologies with which we might explore it, for there to be lots of different kinds of pursuits.

We haven't moved in that direction as readily as we might. I would like to see psychology have more of a connection to what goes on in the world. That doesn't mean that more of it should just be "applied." In fact, many of the attempts that go on within the discipline that seek to have some kind of impact on the world are carried out in a problematically instrumental way. And the values that undergird those efforts are often either unarticulated or under-articulated. I think there's a lot of room for making those kinds of value assumptions more central and having the people who hold them be more self-aware. It's not easy for psychology to talk about the "good," much less the "holy" or the "sacred" or the "divine." That's problematic in its own right. Consider the fact that the first significant dialogue in recent years that's emerged about ethics has come out in the context of torture. What else can we think about when it comes to the ethical realm?

One of the banners that psychology has flown under for many years has been its pride in value neutrality: we deal with what is and we let questions about what ought to be left to others. I understand that; in some contexts, I think it's appropriate. But for me, there is no separating what a human being is from the moral dimension. I think a discipline like our own that continues to sever that connection winds up moving in either amoral or immoral directions.

It is worth noting that there are lots of people doing lots of different kinds of work that bear upon moral and ethical life. Ken and Mary Gergen are engaging constructionist thinking and relational thinking and looking toward broad social issues in order to figure out how we might build a better, more peaceable world. There are people also like Blaine Fowers in our own Division 24 and Frank Richardson and a number of others, Brent Slife, Jack Martin, Jeff Sugarman, people who are deeply involved in these issues in lots of different ways.

Some of that's happening even in positive psychology. This isn't the group to defend positive psychology to because in many ways it's a too simple movement, but some of the aspirations that motivate some of the positive psychologists are not entirely alien to the ones that motivate people in our division. We have more contact with philosophy and we have ways of formulating things that many of us believe are more adequate to the issues. But, I think there's an impulse there and that impulse has something to do with human flourishing and what it might mean to do good work. I think we have way farther to go in this area. The fact that we're able to talk about things like human flourishing is important. I think, and I'm sure many of you would agree with this, there's been too quick a movement to couch that in terms of happiness and to locate flourishing basically in terms of subjective feeling. I think many a psychologist wouldn't have any idea what else to do to address those issues than to locate it in some form or other of subjective experience. That's not unimportant, but I think it's insufficient.

I also think that it's important to think about what the good [i.e., the Good] might be. I'll simply confess to you that, for me, it's not just a value judgment. It's not something that we only arrive at through relation-building and consensus and so forth. I think Good is something that on some level exists in the world. I'm interested in drawing on thinkers like Plato to help me do that. Plato's out of fashion; he's often called an absolutist and all kinds of things, and there are aspects of his texts, *The Republic* (2012) especially, that would lead one in that direction. But there's much of what gets said in terms of the transcendent nature of goodness that I find to be compelling.

I've mentioned Iris Murdoch a couple of times, and she, for me, is an extremely important thinker, partly for strategic reasons. Seen from one angle, Murdoch is not a believer at all; she herself was adamant about not being an explicitly religious thinker. But she's one who still takes on the question of the Good in a way that's very, very important. Indeed, she takes on the question of transcendence. At one point—well, at many points in her work, actually—she has to ask (I'm paraphrasing again), "So what are we talking about here? Are we just talking about a socially valued experience? Is goodness something that we simply agree upon? Is it a value tag that we somehow affix to certain phenomena in the world? Or is there some sense in which we actually are able to discern what goodness is?"

Her answer to this last question is "Yes," and she basically says we engage in this sort of discernment all the time. This doesn't mean that we can contain it. It doesn't mean that we're able to define it in some simple, unassailable way. It means that we're constantly in the midst of encountering a world that's filled with gradations, and these are gradations that we know. There's a sentence in her book *The Sovereignty of Good* (1970) where she reminds us that ordinary people don't think that they invent good. They encounter it in the world, in the plants that they cultivate, in the people that they love, and also when they encounter degraded versions of it in their relationships or in the world. There's a tendency to think that all of those are just human inventions. Undoubtedly it's the case that many of them are. But Murdoch wants to preserve a space for thinking about the idea of the good and the idea of the transcendent, even the idea of perfection, in a way that preserves those ideas. And again, she's able to do it without lapsing into a premature theological claim, as she sometimes puts it, which can be problematic in its own right.

So, there's a lot of interesting thinking going on. I've referred a lot to our division partly because so many of the people are here. But beyond this, we ought to be building a discipline that is able to be in true dialogue with literature and philosophy so that we can learn from people other than psychologists, right? So, that would be part of my platform. I don't know what I would have on the campaign buttons. But it ain't gonna happen, folks! No worries.

DG: I am thinking back to some of your early work that addresses postmodernism and the ethnographic psychological study that you engaged in surrounding some of the cultural movements at the time. It is fascinating. Would you speak to this and help us see how it links to some of your current work?

MF: It's very much a part of my trajectory. And it's odd, in terms of the kind of space that it occupies. One of the projects that I was involved in when I was at the University of Chicago was with the psychologist Mihaly Csikszentmihalyi. It wasn't so much his work on "flow," which some of you may know, but he had been involved in a project that looked at students who had been schooled at the School of the Art Institute of Chicago in the mid '60s. They got a great big grant during the time I was at Chicago to go find those people 20–25 years later and to figure out what they had and hadn't done in the art world, what had been formative and so on.

I somewhat naively had imagined that I would do something on aesthetic development; that is, I would explore questions like: how do people grow as artists, and how do they refashion themselves and come to do the work that most embodies their own self- realization as artists? Those stories were there. But I became particularly interested in the way in which there was often a significant gap between the kind of discourse that many of these people had come to subscribe to—postmodernist discourse, poststructuralist discourse, and so forth—and the nature of their goals and aspirations as artists (Freeman 1994a, 1994b). Many of them became postmodern artists for a time, and they were interested in many of the same issues taken up in Louis Sass's (1992)

Madness and Modernism—issues concerning what can truly be said about the world and the limits of representation. But a number of them wound up encountering these very, very painful dead ends. There was one woman, in particular, who I remember speaking to at her loft in Soho who said, "If I keep on reducing everything, I will eventually have nothing at all." Other people came to feel that some of what they were doing was a kind of charade; they came to feel that some of what they were doing was just inauthentic and inadequate.

A number of them wound up describing a number of different "syndromes," as I referred to them. I called them things like "dissimulation" and "derealization." So, some of the work that I was doing at that point was in what might be termed the *pathologies* of postmodern art. One piece was called "Modernists at heart: Postmodern artistic breakdowns and the question of identity" (Freeman 2000a). And it had to do with those people who decided, in a sense, that all of that deconstructive discourse be damned; they were going to do what they loved and what they thought was more in keeping with their artistic hearts. In its own way, that work also was about the Other—not so much the human other, but about the bountifulness of the world and the possible bountifulness of works of art, which they were both attracted to and repelled by, partly because the language that had the most currency rendered those kind of desires problematic, even retrogressive.

So, yes, I see that work as continuous. I don't do much work in that area anymore. It's because in order to do adequate work in the psychology of art, you have to know a lot about art history and art movements and art criticism and what's happening in *Artforum* and *ARTnews*. I couldn't devote myself in the way that was required by that pursuit. I love it still; I occasionally teach a course in psychology of art and creativity. But it's not as focal as it once was. It's curious. I mean, I did a book on it and it's a book . . . that even some of my closest colleagues have no idea about. It's called *Finding the Muse*, and it was subtitled *A Sociopsychological Inquiry into the Conditions of Artistic Creativity* (Freeman 1994a). Essentially, it was a cultural psychology of art. I should have called it that because I probably would have sold more books! But I didn't know it was that until afterwards. I basically said: you can't look at creativity outside of the context in which it occurs. People were studying creativity mainly by looking at experiments in laboratories; again, that has its own sphere of validity. But to really know about artmaking, you have to know about the market and art history and the culture of art and the nature of the myth of the modern artist, all that sort of stuff. It was a full time job. That was a long answer.

DG: In recent years, you have been writing and speaking about several new ideas (transcendence, creativity in the generation of theory, etc.). If you were to tackle another book in the coming year or two, what would it be about? What are some of the intersection points that you'd be bringing together?

MF: I would probably identify three projects. One has to do with what I earlier referred to as a little element of disingenuousness in my continuing to want to trot out the language of science. The language of science is absolutely appropriate for a significant portion of psychology. So, what I would propose is in no way an attempt to be anti-scientific. That's counterproductive and there's a lot of good that comes about through science. But I have become convinced that some of my own work tries to make the conception of science so elastic as to lose some of its meaning.

I would like to see a portion of psychology more readily avow that it need not *be* science (e.g., Freeman 2015). Much of the work we do is not about hypotheses and "explanation" in the way science tends to conceive of them. It might instead have as its goals compassion, human solidarity, or being able to foster in people some kind of deeper connection to what exists in the world, human and non-human. We could call all of that science, I suppose, but some of it's really not. I mean, when we're doing psychology in the course of our lives, when we're understanding people, when we're understanding ourselves, when we're trying to figure out what makes people tick and what makes them not, I suppose one could say that we're always working with lay hypotheses and the like. But I'm not even sure that that's true or necessary. Psychology has always had an ambivalent relationship to the humanities, to literature and art and so on. There's a division here and a division there. But look, where do we find some of the most compelling and, dare I say, *truthful* renditions of humanity? It's in great works of literature and art. For me, the idea of having a portion of psychology try to move in that direction and to create work that actually might have the immediacy, the power to move us, that works of literature have.... Why not? Why *wouldn't* we do that? Why would we leave all of that to others? This partly means being in a more friendly relationship with the humanities. But it also means taking on a bit more of that for ourselves.

A more concrete project has to do with somehow building a significant bridge between my work on narrative and my work on the Other. I've begun to do that, but that's a hard project that may well.... On some level it'll take me the rest of my life. It has to, you know? The work on narrative is about studying people's lives and life stories, and in many ways the central category has been the self. As I continued my work on the self, though, I began to think about selfhood in more ex-centric terms. Even in *Rewriting the Self* (1993), when I dealt with St. Augustine and what development might mean and what are the ends of human development, I was already making contact with the idea of the Other in some ways. I just didn't have an adequate language for it. That was a project that lasted for a couple of decades, and hovering in the background of it were these preoccupations with the Other and otherness, both in terms of the ends, or the *teloi*, of human experience, but also our relatedness to the world. I want to bring those together in some meaningful way. As I say, I've begun to do it, but there are languages that sometimes clash and I know in my gut at this point, if not fully in my head, that there's a way of working it out. In *Hindsight* (2010), too, I talked about the transcendent horizon of life, that kind of thing. But those attempts are not adequate, quite honestly. Those are fledgling attempts at doing something new, doing something different. So that's definitely another project.

I also have a quite personal project. I've mentioned in a couple of different sessions that I have a 92-year old mother who has dementia, and she's in a wheelchair, she's virtually blind, she has chronic obstructive pulmonary disease, and so forth and so on. She was diagnosed maybe seven or eight years ago. She lives in Worcester. I'm very close to her— I've been a part of her life and she a part of mine. We've been in relation, if we want to put it that way, in a big way for a number of years. Now, for the most part, I've just been her son, but I'm also fascinated by what's gone on over these seven years and what I was able to learn about memory, identity, selfhood, otherness, transcendence, and more through her life. I've written several chapters on her (Freeman 2008a, 2008b, 2009b, 2014a). In fact, what I have done is write a chapter for each of the "chapters" of her life over the course of the seven years. I will not complete a book on her while she's alive. But, I'm going to do that at some point. That also is going to be a strange and difficult undertaking because I don't want it to be purely autobiographical; I don't want it to be yet another story of "My Failing Mother" or whatever. In fact, I'm almost certainly going to call it a "tragicomedy." I also want to be able to use what I know about all those different phenomena to create a picture that is at once a vehicle for understanding, a vehicle for compassion, and so on.

So, that's where I'm heading in terms of my own work at this point. It's daunting; each of those projects feels very large. I guess it's because each of them is. But you know as well as I do, this is the stuff that sustains us. I mean, we're real lucky to be able to do all this, don't you think?

DG: Absolutely, absolutely. Thank you, Mark. Moving in an entirely different direction, but in keeping with our conversation about a kind of otherness that people encounter in aesthetic experiences, you are a music buff and I'm hoping you can tell us more about how this integrates with what you've been sharing with us here.

MF: Whew, boy.

DG: This should give us a sense of your prophetic vision for the field!

MF: I'm a rocker. That's probably the simplest thing to say. I mean, I have a full and bountiful shelf of blues and jazz and classical and so forth. But, if I really want to soar, I'll probably put on David Bowie or something else equally powerful, out there. That's sort of where I am, and it might be a function of the life I've been leading these days, where I feel like it's important to be jolted out

> of things. Sometimes, a beautiful piece by Brahms or Mozart will do that. But more often, I want to put on my headphones, crank it up really loud and rock out.

I've had a lot of favorites through the years. In the earlier years, it was anything from the Beatles to . . . You know, the first concert I ever went to was the Doors while Jim Morrison was alive. I saw the Allman Brothers when Duane was alive. So, that's a big part of my youth. I was a singer for a number of years in a band, a rock band. I'd love to do it before I move on, if possible. But, you know, it's kind of like the art project. There's not so much time for things. But that is really where I am.

Actually, I wrote a piece a little while ago on Keith Richards (of Rolling Stones fame) (Freeman 2013). I read Keith Richards' autobiography. It's part of my narrative work. I love reading good memoirs, autobiographies, and so forth. And there was a volume that I had been asked to contribute a chapter to on the idea of persona. I wound up doing a piece called "Axes of Identity." In it, I identified two different axes. One had to do with temporality and one had to do with otherness. Within temporality, I wanted to be able to think about how it is that past, present and future, or memory, acting and imagining, somehow intersected with relatedness to the world of things, relatedness to the world of others, and relatedness to the transcendent. In many ways, it was my first attempt to try to bring together the work on narrative that I've been thinking about for so many years and the work on otherness, and how to do that in a rockin' way. And there I had Keith Richards, who gave me a wonderful opportunity to do it, not just because it's interesting, but because our images of Keith Richards, while at least partially true—you know, he's a derelict and he snorted his father's ashes and all the rest of it—are countered by the fact that he has so many wonderful things to say about the inspiration he received from the blues and the inspiration he gets from other people when he's playing music. This piece allowed me, temporarily at least, to bring things together in a way that was new and exciting and that made contact with a figure that I'm interested in.

So, you know, some of the stuff I do, like some of the stuff we all do, is bound to be kind of dry and academic. But I try to break out of that when I can.

DG: Similar to your passion for music, tell us about your passion for literature and how it informs your work. What would be your hope for the psychological discipline if it were to be more open to literature?

MF: Well, here I would go back to the very first book I did, *Rewriting the Self: History, Memory and Narrative* (1993). What I decided at that point—actually, it wasn't so much a decision; it's just what I did—was to have each chapter be an attempt to work through a particular text. Each of those texts was being employed to illuminate something. For instance, when I dealt with (St. Augustine's) *Confessions* (397/1980), which was the first of these chapters, I was especially interested in how it is that Augustine's life post-conversion provoked him to return to his previous experience and see it anew. For me, it was the classic example of rewriting the self; the triad of history, memory and narrative came alive in that. I was interested in the problem of development in that chapter too.

I was also fascinated at the time by Helen Keller's (1902/1988) autobiography called *The Story of My Life*, a book that you could read if you're in eighth grade but that has extraordinary philosophical profundity to it, although she probably didn't know that. She's famous, of course, for the scene at the well when she discovers the miracle of language. But she also runs into some significant difficulties, partly because at the tender age of 12 she gets accused of plagiarism. She denies it vehemently. It turns out that she cannot distinguish her words from those of other people. And that's because, she says, "everything I read becomes part of the substance and texture of my mind." She has quite specific reasons for why that particular problem became acute, but of course that situation isn't appreciably different from any of ours. Whose words, I had asked at the time, do we speak and write? What does it mean to originate something? How constitutive is language? What is the relationship between language, memory and identity?

I also turned to Philip Roth (1988), his book, *The Facts: A Novelist's Autobiography*, to begin to understand the relationship between the factual and the fictional. In that chapter, I was looking at the degree to which any narrative is fictive (insofar as it involves making, *poiesis*) but still wanting to maintain some kind of distinction between the fictive and the fictional.

For me, all of those books were profoundly interesting to read and to work through. I have a number of favorite texts. As some of you know, Tolstoy's (1886/1960) *The Death of Ivan Ilych* is certainly one. But there are numerous others as well. Why do I turn to them? Because I love reading them and because they often serve to illuminate aspects of the world, the human world, that I want to think about and they do that in a much more compelling way than most psychological studies.

It's one reason that some of my colleagues can't easily distinguish what I do.... It's why they gave me that weird report, you know? What makes you a psychologist rather than a literary critic? Probably nothing; I don't know. So, all of that is real important and will continue to inform in some way everything I do. I mean, the work on narrative and self and identity and memory and language... that's another lifelong project. Right now, it's receded a bit as I turn to other things. But it'll always be there because it always *is* there, right?

DG: Mark, it's such a pleasure to hear your thoughts and to cover such a range of topics related to your work and your vision for the field. You know me well enough to know that I, as a young scholar, I'm very embarrassed about the state of our discipline. I fear that psychology is finding itself in some dangerous places that lead many clinicians, theoreticians, researchers, and scholars to question the viability of the profession. Your work actually gives me hope. Your mentorship has given me courage. And I see you not as a warrior who's been knocking at APA's gates, but rather a poetic being who calls for richer language, richer theories, and richer approaches to human life. Thank you so much for the gift of your work.

Bibliography

Augustine, St. (1980). *Confessions*. New York: Penguin. (Original work written 397)

Brooks, D. (2016). *The road to character*. New York: Random House.

Freeman, M. (1984). History, narrative, and life-span developmental knowledge. *Human Development, 27*, 1–19.

Freeman, M. (1985a). Paul Ricoeur on interpretation: The model of the text and the idea of development. *Human Development, 28*, 295–312.

Freeman, M. (1985b). Psychoanalytic narration and the problem of historical knowledge. *Psychoanalysis and Contemporary Thought, 8*, 133–182.

Freeman, M. (1989). Between the "science" and the "art" of interpretation: Freud's method of interpreting dreams. *Psychoanalytic Psychology, 6*, 293–308.

Freeman, M. (1993). *Rewriting the self: History, memory, narrative*. London, UK: Routledge.

Freeman, M. (1994a). *Finding the muse: A sociopsychological inquiry into the conditions of artistic creativity*. New York, NY: Cambridge University Press.

Freeman, M. (1994b). What aesthetic development is not: An inquiry into the pathologies of postmodern creation. In M. Franklin & B. Kaplan (Eds.), *Development and the arts: Critical perspectives* (pp. 145–164). Hillsdale, NJ: Lawrence Erlbaum Associates.

Freeman, M. (1999). Life narratives, the poetics of selfhood, and the redefinition of psychological theory. In W. Maiers, B. Bayer, B. Esgalhado, R. Jorna, & E. Schraube (Eds.), *Challenges to theoretical psychology* (pp. 245–250). North York, ON: Captus.

Freeman, M. (2000a). Modernists at heart? Postmodern artistic breakdowns and the question of identity. In D. Fee (Ed.), *Pathology and the postmodern: Mental illness as discourse and experience* (pp. 116–140). Beverly Hills, CA: Sage.

Freeman, M. (2000b). Theory beyond theory. *Theory & Psychology, 10*, 71–77.

Freeman, M. (2002). The burden of truth: Psychoanalytic *poiesis* and narrative understanding. In W. Patterson (Ed.), *Strategic narrative: New perspectives on the power of personal and cultural stories* (pp. 9–27). Lanham, MD: Lexington Books.

Freeman, M. (2004). The priority of the other: Mysticism's challenge to the legacy of the self. In J. Belzen & A. Geels (Eds.), *Mysticism: A variety of psychological approaches* (pp. 213–234). Amsterdam, The Netherlands: Rodopi.

Freeman, M. (2007a). Wissenschaft und Narration [Science and story], *Journal für Psychologie, 15*(2), Retrieved October 25, 2007, from https://www.journal-fuer-psychologie.de/index.php/jfp/article/view/128/120.

Freeman, M. (2007b). Psychoanalysis, narrative psychology, and the meaning of "science". *Psychoanalytic Inquiry, 27*, 583–601.

Freeman, M. (2008a). Beyond narrative: Dementia's tragic promise. In L.-C. Hyden & J. Brockmeier (Eds.), *Health, illness, and culture: Broken narratives* (pp. 169–184). London, UK: Routledge.

Freeman, M. (2008b). Life without narrative? Autobiography, dementia, and the nature of the real. In G. O. Mazur (Ed.), *Thirty year commemoration to the life of A. R. Luria* (pp. 129–144). New York, NY: Semenko Foundation.

Freeman, M. (2009a). The personal and beyond: Simone Weil and the necessity/limits of biography. In J. A. Belzen & A. Geels (Eds.), *Autobiography and the psychological study of religious lives* (pp. 187–207). Amsterdam, The Netherlands: Rodopi.

Freeman, M. (2009b). The stubborn myth of identity: Dementia, memory, and the narrative unconscious, *Journal of Family Life, 1*, Retrieved March 19, 2009, from http://www.journaloffamilylife.org/mythofidentity.

Freeman, M. (2010). *Hindsight: The promise and peril of looking backward*. New York, NY: Oxford University Press.

Freeman, M. (2011). Toward poetic science. *Integrative Psychological and Behavioral Science, 45*, 389–396. doi: 10.1007/s12124–011–9171-x.

Freeman, M. (2012). Thinking and being otherwise: Aesthetics, ethics, erotics. *Journal of Theoretical and Philosophical Psychology, 32*, 196–208.

Freeman, M. (2013). Axes of identity: Persona, perspective, and the meaning of (Keith Richards's) *Life*. In C. Holler & M. Klepper (Eds.), *Rethinking narrative identity: Persona and perspective* (pp. 49–68). Amsterdam, The Netherlands: John Benjamins.

Freeman, M. (2014a). From absence to presence: Finding mother, ever again. In J. Wyatt & T. Adams (Eds.), *On (writing) families: Autoethnographies of presence and absence, love and loss* (pp. 49–56). Rotterdam, The Netherlands: Sense Publishers.

Freeman, M. (2014b). Listening to the claims of experience: Psychology and the question of transcendence. *Pastoral Psychology, 63*, 323–337.

Freeman, M. (2014c). *The priority of the other: Thinking and living beyond the self.* New York, NY: Oxford University Press.

Freeman, M. (2014d). Qualitative inquiry and the self-realization of psychological science. *Qualitative Inquiry, 20*, 119–126.

Freeman, M. (2015). Narrative psychology as science and as art. In J. Valsiner, G. Marsico, N. Chaudhary, T. Sato, & V. Dazzani (Eds.), *Psychology as the science of human being: The Yokohama Manifesto* (pp. 349–364). Switzerland: Springer.

Freeman, M., & Larson, R. (1986). L'esperienza immediate nella ricostruzione della storia di vita [Immediate experience and the construction of life history]. In F. Massimini & P. Inghilleri (Eds.), *L'esperienza quotidiana: Teoria e metodo d'analisi* [Everyday life: Theory and method of analysis] (pp. 437–447). Milan, Italy: Franco Angeli Editore.

Freeman, M., Csikszentmihalyi, M., & Larson, R. (1986). Adolescence and its recollection: Toward an interpretive model of development. *Merrill-Palmer Quarterly, 32*, 167–185.

Hood, B. (2013). *The self illusion: How the social brain creates identity*. UK: Oxford University Press.

Keller, H. (1988). *The story of my life*. New York: New American Library. (Originally published 1902)

Murdoch, I. (1970). *The Sovereignty of Good.* New York and London: Routledge.

Plato. (2012). *The republic* (C. Rowe Trans.). New York: Penguin Classics.

Roth, P. (1988). *The facts: A novelist's autobiography*. New York: Farrar, Straus, & Giroux.

Sass, L. (1992). *Madness and modernism: Insanity in light of modern art, literature and thought.* New York: Basic Books.

Tolstoy, L. (1960). *The death of Ivan Ilych and other stories* (A. Maude Trans.). New York: New American Library. (Originally published 1886)

Wegner, D. (2003). *The illusion of conscious will.* Cambridge, MA: MIT University Press.

David Goodman is the Associate Dean of Academic Affairs and Advising at the Woods College of Advancing Studies at Boston College, Associate Professor of the Practice in the Department of Philosophy in Boston College's Morrissey College of Arts and Sciences, the Director of Psychology and the Other, a faculty member at the Massachusetts Institute for Psychoanalysis, and Teaching Associate at Harvard Medical School/Cambridge Hospital. Dr. Goodman has written over a dozen articles on continental philosophy, Jewish thought, social justice, and psychotherapy. Dr. Goodman currently serves as the Series Editor for the Psychology and the Other Book Series with Routledge. He has authored and edited several books including *The Demanded Self: Levinasian Ethics and*

Identity in Psychology (with Duquesne University Press, 2012), *Psychology and the Other: A Dialogue at the Crossroad of an Emerging Field* (with Mark Freeman and Oxford University Press, 2015), *The Ethical Turn: Otherness and Subjectivity in Contemporary Psychoanalysis* (with Eric Severson and Routledge, 2016), *In the Wake of Trauma: Psychology and Philosophy for the Suffering Other* (with Eric Severson and Brian Becker and Duquesne University Press, 2016), *The Road to the Living God: Ana María Rizzuto and the Psychoanalysis of Religion* (with Martha Reineke and Rowman & Littlefield, in press), and *Memories and Monsters* (with Eric Severson and Routledge, in press). Dr. Goodman is also a licensed clinical psychologist and has a private practice in Cambridge, Massachusetts.

Critique, Construction, and Co-creation: A Conversation with Kenneth Gergen (Interviewed by Mark Freeman, College of the Holy Cross)

In 1967, Ken Gergen took the position as chair of the Department of Psychology at Swarthmore College. He is now Research Professor at Swarthmore. At other times, Ken has served as a visiting professor at the University of Heidelberg, the University of Marburg, the Sorbonne, and at the University of Rome, Kyoto, and other places. In an attempt to link his academic work to societal practices, Ken collaborated with colleagues to create the Taos Institute in 1996.

Ken Gergen has won numerous awards. They include research grants from the National Science Foundation, the Deutsche Forschungsgemeinschaft, and the Barra Foundation and additional awards from the American Psychological Association, the National Communication Association, Constructivist Psychology Network, the University of Buenos Aires, and Adolfo Ibanez University in Santiago. He's also received fellowships from the

K. Gergen (✉)
Swarthmore College, Swarthmore, PA, USA
e-mail: kgergen1@Swarthmore.edu

H. Macdonald et al. (eds.), *Dialogues at the Edge of American Psychological Discourse*, Palgrave Studies in the Theory and History of Psychology, DOI 10.1057/978-1-137-59096-1_7

Guggenheim Foundation, the Fulbright Foundation, and the Alexander Humboldt Foundation.

Ken has also written a number of important and influential books. *The Saturated Self* (1991), subtitled *Dilemmas of Identity in Contemporary Life* (1991) is one; among numerous others, there's also *Realities and Relationships: Soundings in Social Construction* (1994); *An Invitation to Social Construction* (1999); and, I believe most recently, *Relational Being: Beyond Self and Community* (2009), which is a book that represents a new chapter in his thinking.

All of these more "official" matters aside for the moment, I want to acknowledge that Ken is a truly inspiring person, with an extraordinarily lively and imaginative mind, a fierce commitment to the good (though he might be reluctant to use that sort of language!), and an admirable willingness to put in the effort that's necessary to attain it. He's also a friend, with whom I have had the great good fortune in recent years of dining and drinking and sharing ideas at lots of terrific places, generally in celebration of something or other tied to our shared efforts on behalf of the Society for Qualitative Inquiry in Psychology (SQIP). I cherish those times, and I'm extremely grateful for them. As I'll acknowledge in greater detail shortly, there remain issues that divide us. It's been that way from the start. What's amazing is that these differences have never touched our friendship. Indeed, they've sometimes strengthened it. So, he's a *good* friend, and it's my great pleasure and privilege to introduce him today.

Mark Freeman (MF): I have to tell you, I feel like I owe a significant portion of my earlier career to you. And I want to say this in a way that conveys as much feeling as I feel. It's not because I was a student of Ken's, and it's not because I was a devotee of social constructionism. On the contrary, as you (Ken) may recall, I found myself somewhat at odds with some of your work, especially when it came to thorny ideas like objectivity, reality, truth, and so on. There were parts of it I could connect to and internalized and they're still with me. And there were parts of it back then that

I wasn't quite sure what to do with. On some level, I'm still not. That's why I'm looking forward to this interview.

Kenneth Gergen (KG): You can have a therapy session.

MF: A therapy session, exactly. What was great about it is that those initial encounters made me *think*. They made me struggle, and they made me write. They provoked me to work out my own angle on things, and for that I really am especially grateful.

I've always admired your work and the tenacity with which you've pursued it. It's unfailingly thoughtful, provocative, sharp, and significant. But I must tell you, what's drawn me closer to you as a person has been our joint endeavor in recent years working on what's now called the Society for Qualitative Inquiry in Psychology.

So it's great to be here with you.

KG: I feel the same.

MF: Good. So let's begin. I thought it might make sense to begin at the beginning, or close to it. First, what is it that that drew you to the discipline of psychology and what sort of work were you most interested in pursuing at the start? I'll ask you to just keep that one in mind.

The second question has to do with what led up to your landmark 1973 paper, which I'm sure some of you know, called "Social Psychology as History." Generally, I want to know what pointed you in that direction. The next part of this question is a social constructionist question: How did you manage to break out of the particular way you had been constructed as a social psychologist? What is that allowed you to somehow see it from afar, see it anew and move it in a different direction?

KG: You know, every time one asks a question like that—where did it come from, what brought it about—it's in a different context, and each time you come up with a slightly different account of what has happened to you.

MF: Right.

KG: Let me try only one possibility. Let me go back even to early childhood. And it's not a matter of my mother and father, but a matter of growing up the son of a university professor. My father was a mathematician, and we lived in an academic community—this was at Duke University. That community was settled in the countryside. So, I went to public schools with a lot of rural kids, many from families with little education. My family lived in a sort of academic—well –

MF: Enclave?

KG: Good, a small enclave that the local people called Goon Hollow. So I lived one life at home and quite another one at school. I had to talk in a certain way with a certain accent and have certain values at home, and become a totally different person at school. There I learned to be "Southern." By the time I finished high school, I was really more southern in my lifestyle than academic northeastern. Okay, so I'm living a split life, residing in two opposing worlds. Actually, there were more worlds than that, but this split is the most dramatic.

Then I was accepted at Yale. And I take a confederate flag to Yale, totally déclassé. Now I've got another split, bringing with me a "self" that doesn't fit yet again. And so I must make another transformation of self. So a lot of my life has been involved with finding myself in contexts in which I don't fit. I've had to reformulate and rework the "self" to new contexts.

So these experiences stimulated a lot of reading on the self, even as an undergraduate. For example, I was deeply into Sartre at the time, and when I went to graduate school, I think in a sense it was issues of the self that most concerned me. Also, the fact that what psychology promised at the time was, for one, an open space for creativity. I mean, all the problems hadn't been solved. It wasn't as if all the major theories were there and you were just there to work out the implications. There was great room for flowering.

It also interested me that here was a science which promised to provide answers to major questions of well being, and to do it systematically with research that could be given away to people to create futures in which we could all flourish. It was a terribly optimistic view.

So, here in psychology I could work on these issues of self—who are we, who am I—and do it systematically, with the hope that the society would ultimately benefit. I could proceed with the sense that there was an open space for creativity. I went into social psychology, which is the closest I could come to finding a field where these issues could be explored. I entered a graduate program that was totally empirical, and which became a center or experimental social psychology. There was great optimism in the experimental method; it was the future. Now, why did I abandon this venture?

In fact, when I think about my dissertation (Gergen, 1965) at the time, I was essentially conducting an experimental study showing how in a dialogue, using one could respond to the other in a way that would change the person's personality—at least within that space. At the same time, artificial situations, statistics, graphs and so on. Wonderful, awful stuff.

MF: Did you find that alienating at the time, or was it just sort of part of the furniture of psychology?

KG: You know, it was part of the furniture, but it was worse than that. Because there was a kind of smugness about it that was shared throughout the experimental community. There was a sense of superiority, that you as an experimenter could manipulate the situation and get people to do different things. In retrospect it was exploitative and alienating. But I was a very earnest guy. That was not the way I saw it at the time.

MF: I still find you to be, by the way.

KG: Manipulative or earnest?

MF: Earnest!

KG: All right. So, there were a couple of other things going on at the time. One was Erving Goffman's writings. I was not too happy with the mechanistic model prevalent in the experimental area. Goffman's writings on the presentation of self in everyday life began to offer an alternative. My thesis advisor Edward Jones

> said, "Let's test Goffman's hypotheses." That sounded okay, but it started me thinking: well wait a minute, there's a totally different feeling about Goffman's approach. Goffman paints a picture of a person who's in continuous motion, presenting a self to affect the social world around him, not a picture of a mechanistic person who just responds to stimuli. It's a wholly different world.

So now I'm faced with a dilemma. If I crawl into a world in which everything I do is a performance for others, then a participant's actions in any experiment could be looked at as performances for the experimenter. "Hey, I'm doing this for you in this situation." As experimenters, then, we are not *causing* behavior—controlling independent variables—but simply inviting certain performances. And you can also see that these performances are cultural appropriate. So as experimenters, we are simply swimming about within cultural mores. If that's so, then this whole idea of accumulating knowledge through experimentation begins to wobble. We are not marching into the future, knowing more and more about human behavior. We are simply following the currents of social change.

So I presented some of these ideas to the Society of Experimental Social Psychology, and the editor of the major journal said, "Hey, there is finally something interesting being presented; why don't you write this up?" That became "Social psychology as history." But I've got to say one more thing.

The question is why did I dare? I was really fortunate enough to have a mentor at Duke University, Sigmund Koch, who was a radical critic in a sense, and proud of it. I mean, Koch would take on any theory and move with it, and move against it, shaking up all the assumptions and logics. I really admired him. So there was probably some emulation. I would model Sigmund Koch, and take the risk. Go ahead and do it; put it out there; see what happens. At the same time, the earnest side takes over again. I also thought I was participating in in a tradition of honest reason and exploration. I believed we were in a scholarly discipline and everyone would simply join in a collective search for a rational analysis of our condition. "I don't have all the answers," I would admit, "but here's a problem for us to chew on. Let's talk about it." But that was totally naïve.

MF: It didn't quite work that way. So, what was the response initially to that piece?

KG: Well, the article became a centerpiece of what was called the "crisis in social psychology." There was a huge, critical reaction, no balanced inquiry at all, and at times it was personal.

MF: But what was the substance of the criticism? Was it that you were anti-experimental or anti-universalist? In other words, what was the big crisis?

KG: You know, that's a good question. I was really rewarded by buying into the experimentalist program, not only in graduate school (where I left already with a half dozen publications), but in managing to get the cream of a crop job at Harvard right out of graduate school. Everyone wanted that position. I was on the National Science Foundation's panels in two years; I had major grants from the National Science and the National Institute of Mental Health; I was on the board of the major journals. I mean, it was tremendous– I was having a very successful career within the guild. So I was a deep insider, and I had plunged a dagger in the heart of friends and colleagues. How could I do that? It was like a Judas within the ranks. So your question is a good one.

I think there was also an existential problem here, because I was part of a tradition that shared a belief in science as a march toward truth, accumulating knowledge over the centuries, and as we contributed our research results to the journals we were making contributions that could ultimately lead to improved life conditions.

I had pulled the rug out from under all that—because I effectively said that what we are doing in these experiments is simply catching culture at a time in history. We are mistakenly looking at culturally and historically embedded actions and treating them as contributions to universal, trans-historical knowledge. The leap from the local to the universal had no warrant. It's like Yale graduate students back in the behaviorist era when they would look at rats going through a maze and cheer, depending on which theory of human nature the rats confirmed.

But I must admit there was another piece of my argument that stuck the knife in just a little further. I was also arguing that as we do science and create ideas about what people are and what they do, and these ideas are fed back into society they can change the society. They become inputs into the cultural ideas on which people base their actions. So, it's not simply that the science fails to accumulate knowledge; it's actually inviting social change, and simultaneously undermining the generality of previous studies.

MF: Was it your perspective back then that there literally was no room at all for the accumulative project? That is, it was inconceivable to you that there were aspects of social psychology that warranted moving in this universal direction? Or was it that social psychology had sort of underplayed history? How far did you go back then?

KG: Yes, there's an unstated issue I was laboring through at the time, and I would address it differently now. But you might say that the natural sciences did seem to accumulate knowledge, while my argument was aimed primarily at the social sciences. The question then becomes whether we need a different philosophy of science for each. Aesthetically this is not a happy outcome. I was reading a lot of philosophy of science at the time, but I didn't quite know how to solve that problem. If you look into the corners, one might say, perhaps there are some actions that are so deep-seated biologically, that you can't fail to do them, even if you wished to. I should add that I'm currently a big critic of the whole neuro movement (Gergen, 2010a).

MF: I tend to be as well.

KG: That's sort of my current enemy. I sometimes ask my neuro friends, is there any action that people do in an experiment that if you told them not to do it, they would be unable to obey your command? And if they could not change their behavior—even let's say for a lot of money—then you're probably dealing with nature as opposed to culture. If you cut an artery and it bleeds, you can't simply decide that you don't want to bleed. But if I could stop that bleeding by saying, "I'm not going to bleed anymore," then I'll view the action as culturally fashioned. I mean, that's just a rough cut, but it doesn't leave much left for neuropsychology.

MF: Right. Let's continue a bit beyond that epoch. I want to continue with a question about narrative. My introduction to your work actually didn't come through the 1973 piece; it came several years later. It was my first year of graduate school at the University of Chicago, and I was taking "Concepts in Human Development" with Bernice Neugarten. You'll be glad to know that you were on her syllabus, and this was another crisis paper, but it was one you had written. It was called, "The Emerging Crisis in Lifespan Psychology," (published in 1980) in which you introduced an "aleatoric" perspective on the developmental process emphasizing chance, accident, even randomness. Aleatoric had been a term used by John Cage in talking about musical compositions.

KG: Right.

MF: That was actually the first piece that provoked me to write a rejoinder of sorts; the emphasis on randomness, I had said at the time, was tied to a perspective on development that looked essentially forward in time. Looking backward, though, engaging in narration, led to a quite different sense of things: a story could be told, one that generally hung together more than the aleatoric perspective had suggested to me. So, I want to turn that question into something, again, that's about you and your life and work. Looking backward at the post-1973 trajectory of your life and work—and I know that's a long swath of time—how would you characterize the story at hand? Is there a storyline that you could trace, or is it haphazard? When you look back, is it a series of sort of completely unpredictable shifts? Or is there something that somehow, dare I say, *evolved* in terms of your own intellectual project?

KG: Well, that's kind of a clever question.

MF: Why, thank you.

KG: You are asking me about whether there is an evolution, as if I could give you an earnest answer. But that's not quite fair, because you invited me into a cultural tradition of telling stories. So I cannot answer earnestly about my evolution without undermining the constructionist assumption about narrative construction.

I mean, to tell a good story—one cannot be random, it's not aleatoric. In a good story, one thing causes another, and events move toward some end or goal. That's part of our way of telling stories. So you brought me in and said, "Hey, let's play that tradition." And I've played it for you. Now, is there something to my trajectory, you ask, a real progression? Sure. If I crawl into that story space, I can get totally carried away with its reality. But there are also multiple stories that could be told. I could probably, if you pushed me, tell you the story of the development of an idiot. I mean, why would I shoot myself in the foot by attacking my guild? What sort of pratt would do that? Because I was basically eliminated from the experimental guild. I mean, all my companions pretty much left me. I was essentially booted out, with nowhere to go; I was kind of a Ronan. I subsequently spent time in anthropology and sociology, in Division 32 with the humanists, and so on. I didn't have a home. Only an idiot would want to do that.

And I could also tell it as a funny story, a clown story. So I'm telling a story here of a certain kind, and it makes sense to me because that's part of what storytelling does. It takes a whole lot of things and makes coherent sense of them. Is there a truth in it? Sure, there's a cultural truth in it because I'm using the implements that we do to make truth for us at a given time.

MF: I need to push a little bit.

KG: Okay.

MF: I mean, it seems to me fair to say that, on some level (and I guess this is my own story, my rendition of your intellectual history), one could look at "Social psychology as history" as being a kind of forerunner of what eventually became certain aspects of social constructionism. It seems to me also that the work that I just cited on this alcatoric perspective also led to the particular version of narrative that you've just voiced. So I'm seeing, I think, some threads? Now indeed, I'm doing the seeing and on some level I suppose one could say I'm creating the story. But there's something *there* in the work that you've done, in the commitments that you've had. And I guess what I want to know is what have been the driving ideas that have characterized this path—however you want to tell the story? If you had to identify the pivotal chapters in Ken Gergen's life and work—and I'm not asking you to name them (I'd originally asked him to do that; he didn't like it!)—what would they be?

KG: Well, we have several issues at stake here.

MF: It's true,

KG: I'm not real fond of "basic drive" questions, as if somehow I could look into the core of my being and discover the well-spring of my actions. So let me take another part of your question that I found interesting.

MF: Sure.

KG: If you take some of the arguments that were in that early paper on social psychology as history, I think you can indeed find assumptions that if you nudge them a bit will lead to social construction. Once I make the argument about what I called "enlightenment effects," that is, scientists can change society by sharing their truths about who we are, I am close to constructionism. It's saying that if I share a discourse with others—my constructions of the world -and they join in this discourse, new forms of life may emerge. This is a center idea for constructionists. For example, if I as psychiatrist inform the society about the nature of "mental illness," and people accept this as true, they may begin to construct themselves in these terms. When faced with problems, they may say, "Oh, it must be depression, or I must be bipolar," and they begin to seek out therapists (Gergen, 2005). Herein lies an entire line of constructionist critique. So right there, you have in that early paper a line of argument central to later constructionist work.

This line of thought becomes much more thoroughly developed with time. Along with a host of scholars in the social studies of science, the history of science, and the sociology of knowledge, for example, one can begin to see constructionism as a theory of knowledge. And, as I have tried to argue, as a theory of knowledge there are enormous advantages over the positivist/empiricist fundamentalism that has so stunted the potentials of the social sciences in the twentieth century (Gergen, 1994b). In constructionism we have as well, a theory of knowledge that makes no attempt to declare itself as true. It shifts out sites from looking at science not in terms of its truth-telling capacities but in its pragmatic contribution to society.

Let me expand here a little in light of a question often addressed to me. "Aren't some descriptions of human action simply more accurate than others?" This seems obvious, doesn't it? For example, cultural anthropology does seem to make truth claims, and to do so in hermeneutic or interpretative ways with which I have an affinity. I have even used some of these claims to support various constructionist arguments—for example on the cultural construction of emotion. Yet, at base, I would not wish to make any claims as to the comparative accuracy of competing accounts of culture.

Even in many corners of cultural anthropology, particularly critical anthropology, they more or less understand that whenever you study a culture you are coming to it with your own language and you are going to thresh your observations through your language. Whatever you take back to the home culture is going to carry a tradition of that language. Your description is not a reflection of what there is, but a creation of what there is in your terms. So I don't look at is as objective, but as a construction of the other. To be sure, one can be more or less correct, but only within a shared perspective.

In my recent writing I have been trying to make a strong case for what I call *reflective pragmatism* (Gergen, 2015e). Let's not accept any truth claim as true in all worlds. There are no grounds for making such a claim. You're always working in a specific language, with specific traditions, assumptions, and so on. But pragmatically, such languages and assumptions may be useful (or not). So yes, you could have an ethnography that would be true for us, because that is what we call XYZ in our culture. Given our cultural agreements, I can also go and see if your report is accurate. They have mating rituals, we might say, because that's what we call those patterns of behavior. And yes, all of us who talk about mating rituals can see them immediately. But that doesn't make such an account true or accurate; it just makes it useful for us in talking about them. Now, the question becomes a matter of reflective pragmatics. Why do we want to say that about them? What hangs on this account, and what are the ideological, political, cultural implications of describing their actions as mating rituals? Why don't we call them love affairs, for example, as opposed to mating rituals? Why construct them in this way as opposed to another? What are the values that are carried in our characterizations? For whom are these characterizations useful, and for what purposes? Who may be harmed? These are the sorts of deliberations invited by a reflective pragmatism.

But now let me return to your earlier question about social psychology as history and its implications for later work. I need to add that the critiques of that paper also stimulated new lines of thought. And some of these new lines became central to later ideas in constructionism. What emerges next in your own dialogue may very well depend on what questions people raise.

MF: Sure. Sure.

KG: I'll give you one important example. The early, and most powerful critique of my arguments essentially said that I was right about historical changes in patterns of social action. These, however, were only superficial. But we as scientists are not interested in superficial change; we are after something deeper, let's say the basic or fundamental processes of cognition, motivation, and so on.

An interesting critique. But then, for me, the question comes up: how can you tap into these fundamentals? How can you infer from the surface behavior—which I'm going to call cultural—that a "fundamental process" has been at work? How would you identify the process? Now, this is where it's going to become interesting for you, because this is also the hermeneutic question. How do I know from your words what private meanings they express? How do I take the narrative account and know what underlying meaning it represents? How do I access your subjectivity? And if you can't answer that question—and nobody can—then why do we presume there are processes, meanings, motives and so on that determine the "surface" of our actions? Hermeneutic theory has wrestled with this question for centuries, with no compelling answer. If you can't answer the question about how one can make a valid interpretation, then what are we doing when we say we understand another's subjectivity? What are we doing when we say I have empathy for that person? Can we make sense of the project that links understanding to our ability to somehow penetrate behind the eyeballs of the other?

As I said, I don't think the hermeneutic question is answerable in principle. And this impasse led me ultimately down the path of post-structuralism (Gergen, 1988). I lost interest in theorizing or researching the structures *behind* the actions. Let's engage with the actions in motion. Now, we could go on with that if you want.

MF: I guess I would question your characterization of the hermeneutic project as you've just articulated it.

KG: Yes, you can do that.

MF: If I'm trying to understand the other, I don't know that the aim is to somehow "reach behind." I don't know that it's an inferential process where the goal is to somehow be able to discern the other's subjectivity . . .

KG: I'm not sure, either.

MF: What I need to do is I need to learn how to be a reader of sorts. And I need to be able, to the extent that I can, to be conscious of my prejudices and to bracket them where it's possible so as to let this other being, text, speaking person, dying person, whatever, be there in her difference, or his difference. So, I think there are certain aspects of construing the hermeneutic project that are themselves fraught, and the way you've represented it, I agree, is problematic.

KG: Yeah, and you know, we could go on with that issue and it might be fun. But you sort of leave off with the Gadamer dilemma of moving beyond my own horizons to some kind of fusion. I find that very romantic. I mean, it's kind of a metaphoric space of mystery. But, if you suspend all your forestructures of understanding, it seems to me you wouldn't have the capacity to understand anything. You're just a blank slate without comprehension.

MF: Sure. I don't think that's where Gadamer's at, but we shouldn't pursue that too much. What needs to happen for Gadamer is that I need to at least have the ability to have my own prejudices or forestructures *displaced* to some degree. That doesn't mean erasure; it doesn't mean I can encounter the world with a blank slate. It means I have to be open enough to the otherness at hand that what I brought to that encounter initially can be corrected. But let me turn this into a question, and it has to do with the very nature of this conference. The idea of the other, psychology and the other. To what extent do you connect to that language? And to what extent not?

KG: Well, let me first of all say that I approach this question from a constructionist standpoint. So I understand that the discourse of psychology and the other comes out of our traditions. We inherit that language. And even if it's a constructed language, I live with it in the same way you do. Just because the discourse does not represent the world as it is, doesn't mean we should abandon it. I should add that from a constructionist perspective difference does not mean division. There are many scholars in this room who differ with me in important respects—even you yourself. However, because I understand that we are all working with constructed worlds—held tenuously together by a "mobile army of metaphors," as Nietzsche would say—there is no sword of truth that will ultimately eliminate the wrong-headed. Difference does not mean the other is "wrong," and thus somehow second-rate or dismissible. Rather, we learn from those who differ; we see other moves in discourse, other values in action; new spaces are opened for relating.

So, returning to the self, I live within the common traditions, but I'm not content to stay there. Because look at what you've done with your question. In the very construction of self and other, you've already created an ontological gap. There's "me" here trying to understand "you" there, each of us living in our own subjectivity. I don't know how to actually open my private world to you; and I don't know when and how your subjectivity is being expressed. So what you've done is to create an ontological gap where there will always be this problem of self and other, and how one subjectivity can ever understand the other. You've created a world of difference, of social atoms. Going back to Democritus, our language creates a world of independent entities.

Now, that's okay but look with all the problems that you create when you put that discourse into play; look at what happens to us:

> "I live in my world, and you live in your world. You are fundamentally separate from me. I don't ever quite understand you. I'll never know quite what you're feeling or thinking. And if we are each independent, I have to take care of myself, don't I? Isn't that the point of life? That's all I can really know. Isn't everyone else taking care of themselves?"

I'm now playing here, but you can see the implications. Our sciences repeat this vision. Here is Freud and the pleasure principle, with the Id searching fundamentally for its own pleasure, and here is reinforcement theory, telling us that fundamentally we are organisms seeking to get the most for the least. It's there as well in microeconomic theory, and in sociobiology with the selfish gene, and so on and so on. You are familiar with Ed Sampson's early critique of self-contained individualism. Well it's the entire individualist ideology that falls out of that atomistic view of individual, private, and separate selves.

Relationships on that account are not fundamental. Rather, you *build* relationships. You have a relationship between two independent beings. Now, that's okay up to a point; it's what we inherit. But I say from a constructionist standpoint, yes, we can live in that language, but we don't have to remain there. What if there were another way of looking at what is fundamental, not you there and me here, but as relational process. If we could see relational process as fundamental, then anything we say about independent selves or psychological process or subjectivities comes out of that relational process (Gergen, 2009). So relational process is the origin, the Ursprung, of all meaning and that meaning includes words like "self" and "other."

If we could develop a discourse of relational process, wouldn't it have more promising ideological, political, and cultural implications; wouldn't it open up new spaces of being? Can I take that one more step?

MF: Sure.

KG: You see, I have some problems with this conference in that respect. But let me take it a little bit further. I mean, there are a lot of books on Levinas, and actually you have done some work on Levinas yourself. But who is the blessed person in that story? Who in the story about accepting the face of the other, being displaced by the other, listening to the other, or embracing the other, is the hero? It's not the other, but the self. "Hey, I accept you." "Hey, I am compassionate." "Hey, I am Jesus Christ, I love everybody." I mean, it's not the person who receives the compassion, the nurturance, the acceptance, or the love who is blessed. The other is faceless in some sense.

MF: [Gasp]

KG: Yeah, I have gone too far in saying faceless. But in this story others are ultimately a means to end of self-fulfillment. "I am moral!" Now, that's a critique, and I'm not saying I would pursue it everywhere all the time. But given this kind of critique—emphasizing self as opposed to other—what if the ethic were built around relational process? Can we ultimately be responsible to a process from which all meaning and all possibility of morality of any kind emerges?

MF: Here I'll be a pluralist. There are plenty of things I might say in response to that characterization. But I really do want to be pluralistic. I also want to affirm the relational. I mean, I don't think it's possible to talk about the other, even in a Levinasian sense, without it being relational through and through. So about that, we're of a piece. But I wonder whether or not you're putting aside the idea of the other maybe a bit too quickly? And I wonder whether there might be –

KG: I think about ten years.

MF: Hmm?

KG: No, but I think about ten years too soon.

MF: Why not say that there are situations in which the language of relationality and relation building really is primary? I know some of the good work that you do at Taos and other places where that language is completely fitting. And I can also think of other situations, some of which I've written about when I go to see my mother, a 92-year old woman with dementia. It's not heroic caregiver stuff. But I like to think it's not about me. It's about her and what it is that she draws forth from me. So in that context, the relational language doesn't do for me what the language of the other does. Is it possible for us to say let's figure out where these languages best work and where they don't?

KG: Yes, absolutely, that seems just right. Again, I don't want the traditional language of self/other to disappear, and it's possible that that kind of language works very well under certain kinds of conditions, in the same way that a scientific language could work in other conditions. For example, you could say "I'm going to see my mother to give her certain kind of treatments with music, which may bring her back to some kind of communication." And within that sphere of construction, you could do it objectively, even measuring the effects of your treatment. So, what I'm after is not to displace our history; it's to add a whole new way of orienting ourselves, opening up new kinds of spaces to think through our lives and institutions.

Here, for example, if we thought of what you're doing as nurturing a relationship or a relational process, I wonder if your actions would change their form? I should add that I have been criticized in my relational theorizing for eliminating human agency. And within relational theory this is indeed the case. However, it's important to realize here that the concept of agency gains its meaning largely by contrast with the concept of determinism. So if you bracket agency, you've also eliminated determinism. The terms feed off each other in their definitions. What I am trying to do in relational theory is take the entire determinism/agency bifurcation, and saying, "Okay, these are discursive traditions. We don't have to fight out whether people are truly agents or determined; these are just two ways of talking, two ways of interpreting. Each has some utility." But what if we bracket the binary, and explore a way of explaining human action that sees it as emerging from within relational process. And of course, you could say that at any point in the process one can "choose not to participate." But let us not look at this "choosing" as an exercise in agency. Choice emerges from the relational process (Gergen, 2009).

Now, one can still counter, that I do remain with an ontology of individual, embodied beings. And there is a way in which this is correct. The very language we employ to create theory demands entification, that is, a world of separate entities. We could scarcely create a theory about human action that did not use nouns and pronouns. Thus, you might say, the attempt is to create a theory of dance, but without separating the dancers from the dance. Now, I can separate out analytically each person's movement. But that movement by itself makes no sense at all. You can't tango alone. It's the dance that counts. Consider what we're doing right now—if you took the words alone, they would make no sense; they make no sense without what you have said. My words are nonsense outside of what you've said and vice versa.

What I suggest in my 2009 book, *Relational Being*, is that we look at our bodies as carrying an enormous array of resources. For example, the language I'm using. I didn't make it up; it doesn't come from an "in here;" it comes from dialogue; it emerges from a relational process. My posture, what I'm wearing . . . it's all emerging from relationships. At the same time, there are only some of these relationships represented in what I'm doing here right now. So, I always have the possibility of shifting from one form of action or performance to another—playing with you, criticizing you, and so on. And I don't want to create a sort of a super-consciousness that selects which performance will emerge. What I want to say, is that we can participate in different kinds of dances with one another, and one or another dance movement will emerge as it becomes salient to the unfolding relationship at the moment. So that if you ask me a question, for example, I'll try to give you an answer, because that's just what we do in our culture. I don't have to answer, but it's so well embedded in our tradition that I'll probably do that. Unless you ask me a question that I don't understand, for example, then I'll say, "Well, tell me more." Or, if the question was curious, I might say, "Why are you asking that question? What hangs on that question for you," or something like that. So I have available many possibilities other than answering the question, but they all come out of our traditions. I don't so much select one freely, as move within the limited space of what is available in the particular relationship.

MF: Right; for sure. Let's continue. I want to ask one more set of questions related to the ideas that we've been considering. Returning to the term "construction," it seems to me that in recent years you've taken that idea in a somewhat different direction. If I look back at your earlier work, a lot of your concerns were epistemological; they were about questioning realism and objectivism, that sort of thing. But much of your work now is about building a better world. I'm thinking especially of the work that you and Mary are doing in terms of the newsletter, the Taos Institute, and so on. You've identified your work as being part of a "future forming" project (Gergen, 2015a). Can you talk a little bit about that?

I do have to ask you one more loaded question too; it's a friendly one. A number of years ago, when I began to see you moving in this direction, I actually suggested to you that there seemed to be almost a *theological* dimension to some of what you were doing. And, in fact, there I saw the title to chapter 12 in your excellent book *Relational Being* called "Approaching the Sacred." What's happening? And what *is* sacred?

KG: That's a lot of lovely questions. All right, let me just touch on a couple of things. For one, that epistemological battle between empiricism or realism and constructionism was of major importance to me for about ten years, at least. Many fights, a lot of scars. I've got Karl Popper saying, "You are the enemy." Things like that.

MF: Do you remember when Louis (Sass) and I were on a panel together, it was called "Postmodernism and its Malcontents"?

KG: Exactly.

MF: We were the malcontents and you were the discussant! It was good, clean fun, but nonetheless.... But go ahead.

KG: But at some point—and I think Mary's the pivotal person here—she says, "Are you going to continue to fight those battles forever?" And she was right. Why do you want to stay in that space? You know practically everything you're going to say and it's just a matter of trying to point out the shortcomings of the competition.

MF: Right, it's a finite project.

KG: It's also finite in the same sense of critical studies, which came out of much the same context. I mean, critical studies are filled with constructionism, critical psychology being a case in point. It's a finite orientation. You deconstruct all the essentialisms—and it's emancipatory. But then what?

So deconstruction emancipates us from traditions, and that's fine. But why not, to go back to the constructionist premise, use language to create something. Construct worlds that create the future, not just emancipate? Why not construct in order to activate? Here I'm being instructed by some friends in a management school who developed a dialogic practice called appreciative inquiry. For example, they're consultants and they are faced with organizations in conflict. But, rather than going in and studying the conflict and telling them what to do, they go in and set in motion a dialogue about what they value. "What's important to us? What do we care about?" That dialogue brings people together in a more productive way—and then they build on the dialogue. They base the practice on constructionist ideas. As they propose, problem talk is only a form of construction, and it doesn't get us very far. The problems become increasingly apparent. Let's begin to talk about common visions and values. Now we have positive transformation.

That was illuminating to me because then you could approach constructionist ideas with the attempt to explore practices that would accomplish something in society. Where do the ideas lead? Now, that's part of the basis of the Taos Institute, which tries to bring theory—mostly constructionist theory—together with practices in therapy, education, peace building and so on. As we often ask, how can we use these ideas to create the future? If we talk together, if we find ways of collaborating, if we find the right mode of dialogue, things can happen, we can bring about change. But this change—and again—this is the relational part, has to be lodged in relational process.

MF: I really do admire that dimension of your project. I have to ask you a question, though, in this context, and I'm not sure what you'll make of it. But what kind of future? I realize that whatever future is built needs to come out of dialogue with people and needs to come out of relation and so on. But I'm sure you would agree that not every future is worth having, that there are some that are perhaps more worth having, and so on. I'm not asking for an absolute or anything of that sort. But what would you say are the sorts of values, aims, purposes, that really support this work? You just mentioned the idea of peacebuilding. What else? How articulated is the vision? Or is it something that's ever on the move?

KG: Let me add a little bit of a footnote to an earlier question, and then move on to this issue. Theology: what do I do about that, you asked? Why that chapter in *Relational Being*?

MF: I like it, by the way.

KG: Oh, great.

MF: And lots of your other work, too.

KG: You are kind. Well, first of all, I have a lot of friends who really have a strong sacred sensibility. Surely that's a constructed world, but every construction may have something it does for us. And for many people the discourse of the sacred has profound consequences. You don't abandon a worldview because it's a construction, because every intelligibility creates something of value for us together. So then I say, well, let's take for example the auratic quality of that discourse—the aura that pervades the discourse of the sacred, and ask ourselves how it can be linked to relational process. Could we speak of relational being in such a way that it acquires a sacred dimension? And can we move away from the conception of a sky God, to prizing forms of relational practice such that we could speak of the sacred dimension of our every-day actions?

MF: But what makes it sacred?

KG: Nothing *makes* anything sacred.

MF: Is love part of this or is that too loaded?

KG: Again, it would depend on what you want to bring into the account. You have to watch very carefully what you're going to objectify. I tend not to want to do the love thing because that word is so dispersed and with so many meanings, and so many look at love as a panacea for everything. "If we could only love one another, everything would be great." You know, we've been through that for about, what–

MF: A long time.

KG: Twenty centuries and we still don't love each other.

MF: I'm not wedded to that particular term. I'm asking you, though, because it seems to me that you're not only committed to relations, but to relations of a specific sort—those that grow people and bring them closer together versus those that don't. I'm trying to figure out in a certain sense where the directives, implicit though they may be, come from.

KG: Okay. But realize I'm not universalizing it. There is a stance that I kind of live with from day to day and I don't know whether it'll be there forever, or under all circumstances. But what I'm trying to place the greatest value on is the well-being of the relational process itself. The process may both unite and divide us, but it's the process that is important. Once you have divided communities, the process is severed or eliminated.

So I've been working with an idea of first and second order values. First order values are always in motion, being created among us in every situation. As we begin to talk together, we're either sustaining some value tradition or creating locally. So that process is always under way. But when these first order moralities become concretized and universalized and become what we call fundamental values, then you're in trouble. Then you begin to draw a circle around who is good and who is not good, and those who aren't in the circle of the good are "treated," or put in prison, or eliminated.

So, you need a second order value, which brings those differences together, a second order ethic that is played out in practice, not in theory. The ethic gains its momentum and meaning from practice, how you do it. Part of this is also a critique of the theoretical project, in which so many of us are involved. We have somehow hoped we could solve issues of value theoretically. And so we have interminable books and articles on philosophic and conceptual issues of ethics. I've done that for a long time.

But, you know, that's a Cartesian vision, that you could somehow set everything aright in some rational fashion. I don't buy that anymore. I think what you've got to do is work with the relational patterns in which we are all engaged. How to do it? How we can move past the adoration of love as an abstract value, for example, and learn how to have loving relationships? What do you say, what do I say, what do we do in actually carrying it out? So I'm very interested in the actions essential to creating and sustaining relationships.

MF: I guess what I would want to ask here, again, is: why not move in a more pluralistic direction? I want to just make quick contact with something I read in the pieces that you sent. In the world-making piece, you ask: What if, "rather than searching for the determinants of depression inquiry, we're launched into means of escaping or avoiding [it]?" What if, "rather than revealing the suffering experience of immigrants, inquiry is directed toward advancing immigrant well being?" And so on. So this is your idea of future forming. But the question I have is, why couch it as a "rather"? In other words, wouldn't it seem important to know something about the possible sources of depression in order to figure out what you're going to do about it? Wouldn't it be valuable to know something about the nature of immigrant suffering in order to spell out what well being might be? So it's not so much a plea for theory, but I get the sense sometimes that you want to jettison that whole what-is project and replace it with something else. I wonder about the replacement idea.

KG: Yeah. You know, I'm always overstating everything partly to make a point.

MF: It works!

KG: Because otherwise, we can just play around the edges forever. I guess it's trying to put a stake in the ground and saying, "Okay, what are we going to do about that?" So, the essay on future forming—which won a prize last year for social science essays –

MF: It's a good essay.

KG: It was trying first to point to the way our research tradition documents everything (Gergen, 2015a). That's what we do. Based on an ocular metaphor, research is about documenting what is the case and putting it into articles that fill up journals that no one reads. It only becomes worse and worse because of demands for more and more publications. Thus you read little, except for locating references needed for your next article.

So you've got thousands of articles coming out of the social sciences that have almost no impact on the society. It's almost useless. That's overstating it, but I could defend it up to a point. What are we doing with our lives, writing in ways no one can read in any case? Only our colleagues can read it, if they must. This is one reason Mary and I often turn to performative social science.

MF: I know. I was always surprised when my parents couldn't read my stuff. I thought it was so accessible.

KG: Yeah. So why do we write in this awful language? And why do we treat others as objects for study? What is that all about? What is this whole notion of "the other" that I am going to study? I am going to document you; I am going to get your narratives and a publication from that. I have problems with the whole process. Again, I don't want to jettison anything, but I do have problems with it.

So then I say, why not shift our challenge from this tradition of documenting with that of creating the future. For example, rather than research attempting to document our understanding of depression—

which we've been fruitlessly doing now for almost a century—why not work on ways to create change. Let's say, we've got a person who's feeling pretty punky and suicidal. How can we talk to that person? What kind of dialogue could we have? Let's establish a relationship with the person, try things out. How did you do it, then, and what kind of relationship did you have? What happened here, and how did it work out there? We could help each other create futures. The point is not to write yet another paper, but to create new and potentially valuable forms of practice.

MF: Well, here's a question that I hadn't anticipated asking—and it's not one I'm unsympathetic to, mind you. Are you essentially calling for the end of social science?

KG: Just one point here. Natural science gains its esteem–

MF: Credibility.

KG: Because of what it produced, whether the electric light, the atomic bomb, a cure for typhoid fever, or whatever. No one gives a whit or even understand the theories—the truth posits. What those sciences do, what makes them culturally significant, is their contribution to people in terms of their everyday lives. So the question is, what have the social sciences contributed? People don't need, for example a thousand new journal articles. What have these contributed to people's lives?

MF: I'm with you.

KG: What they do need are new practices of how to love.

MF: All right.

KG: So, that—I don't want to give up the tradition, I just want to shift the focus to forming the future as opposed to documenting a past that rapidly disappears.

MF: I'm not sure how much time we have; I hope we have time for one or two other questions. A quick comment on the Society for Qualitative Research in Psychology.

KG: Okay.

MF: Why were you as invested in it as you were? As you are?

KG: As I am.

MF: Which we're all glad for.

KG: Qualitative research in psychology– we spent seven years working to get qualitative research recognized by the American Psychological Association as legitimate, as something worth doing. Yet, I don't do qualitative research.

MF: For the most part, I don't, either.

KG: I mean, Mary and I do performance work, whether you call that research or not. We do. But for the most part, you know, I champion the qualitative movement because it represents a pluralism (Gergen and Gergen, 2012). All those practices included in the qualitative movement come out of different traditions. Phenomenology comes out of a different tradition than most of narrative studies, and that differs in basic assumptions from discourse analysis, which can also be contrasted with action research, and so on. They each have theories, assumptions, values, perspectives, constructions, that together enrich the field enormously. They open a space for an enormous range of activity outside of testing hypothesis. Look at all the things we can do. For me, that kind of enrichment was first of all significant.

But you might say that more privately, the pluralism is also a constructionist venture. If we appreciate that there are many different paradigms, each with its own potentials, each offering potentially valuable resources, that's an implicit constructionist orientation to knowledge. So, the movement really becomes important in terms of what I think are the ideological political consequences of a constructionist perspective.

MF: All right, good. There's so much more we could talk about in that context. And again, let me just reiterate what a rich and important collaboration that's been. The final question I'll ask before we open it up just has to do with . . . it's probably not the kind of question you like . . . but the Ken Gergen legacy. I know, I know. But, I must say, I did get the sense through the work on SQIP that that's part of it, that there are certain things that you want to leave to our discipline, which I know you still see as fraught and troubled, to some extent. So if you could just say a few words about what it is you'd like to leave the discipline with?

KG: Okay. The problem is that I *alone* don't lead the discipline with anything –

MF: Of course.

KG: Whatever I've done has been part of a group of people who talk and support and help each other.

MF: Absolutely.

KG: So I don't make any claim to any of that. I look at social construction not as mine, for example, but an international dialogue on the nature of truth, objectivity and so on. I'm a part of it, I articulate it in a certain way, but it's, you know, it's not my legacy.

Relational theory is a major step for me in terms of intellectual work. But again, I'm part of the poststructuralist dialogues, not very far from the core of that work. The Taos Institute: I've been very much involved with the development of this institute, which now has 500 associates in 38 countries doing practical things.

MF: That's amazing.

KG: Like changing the Chinese school system, things like that. I mean, for me these are like flowers. All these people making all these fascinating changes.

MF: It's beautiful. And Mary, again, has been a big part of that, I know.

KG: Yes, a big part of it. But with other colleagues as well.

MF: Of course, of course.

KG: The Society for Qualitative Inquiry, to which your contribution was enormous. Yes. *Theory in Psychology,* the journal, was launched with some other friends. The same for *Qualitative Psychology.* I also helped to launch a network on dialogue in Latin America. I mean, it's like you help projects to develop. You're part of them, but it's not your own legacy; it's sort of a legacy of *with.*

MF: Great. Did we miss anything that we ought to have talked about?

KG: Without doubt.

MF: There's plenty more, but let's stop here.

Bibliography

Anderson, H., Cooperrider, D., Gergen, K. J., Gergen, M., & Whitney, D. (2001). *The appreciative organization*. Cleveland, OH: Taos Institute Press.

Bauer, R. A., & Gergen, K. J. (Eds.). (1968). *The study of policy formation*. New York, NY: Free Press.

Gergen, K. J. (1965). Interaction goals and personalistic feedback as factors affecting the presentations of self. *Journal of Personality and Social Psychology, 1*, 413–424.

Gergen, K. J. (1969). *The psychology of behavior exchange*. Reading, MA: Addison-Wesley.

Gergen, K. J. (1971). *The concept of self.* New York, NY: Holt, Rinehart & Winston.

Gergen, K. J. (1973). Social psychology as history. *Journal of Personality and Social Psychology, 26*, 309–320.

Gergen, K. J. (1978a). Experimentation in social psychology: A reappraisal. *European Journal of Social Psychology, 8*, 507–527.

Gergen, K. J. (1978b). Toward generative theory. *Journal of Personality and Social Psychology, 36*, 1344–1360.

Gergen, K.J. (1980). The emerging crisis in life-span developmental theory. Behavior exchange in cross-cultural perspective. In P. Baltes & O. Brim (Eds.), *Life-span development and behavior* V. 3. New York: Academic Press.

Gergen, K. J. (1982). *Toward transformation in social knowledge*. New York, NY: Springer-Verlag, 1982.

Gergen, K. J. (1985a). The social constructionist movement in modern psychology. *American Psychologist, 40*, 266–275.

Gergen, K. J. (1985b). Social psychology and the phoenix of unreality. In S. Koch & D. Leary (Eds.), *A century of psychology as science* (pp. 528–557). New York, NY: McGraw-Hill.

Gergen, K. J. (1988). If persons are texts. In S. B. Messer, L. A. Sass, & R. L. Woolfolk (Eds.), *Hermeneutics and psychological theory* (pp. 28–51). New Brunswick, NJ: Rutgers University Press.

Gergen, K. J. (1989a). Social psychology and the wrong revolution. *European Journal of Social Psychology, 19*, 463–484.

Gergen, K. J. (1989b). Toward self as relationship. In T. Honness & K. Yardley (Eds.), *Self & identity: Psychosocial perspectives* (pp. 53–64). London, UK: Wiley.

Gergen, K. J. (1990). Metaphors of the social world. In D. Leary (Ed.), *Metaphors in the history of psychology* (pp. 267–299). Cambridge, UK: Cambridge University Press.

Gergen, K. J. (1991). *The saturated self: Dilemmas of identity in contemporary life*. New York, NY: Basic Books.

Gergen, K. J. (1992a). The mechanical self and the rhetoric of reality. *Annals of Scholarship*, 9, 87–109.

Gergen, K. J. (1992b). Organization theory in the postmodern era. In M. Reed & M. Hughes (Eds.), *Rethinking organization* (pp. 207–226). London, UK: Sage.

Gergen, K. J. (1994a). The limits of pure critique. In H. Simons & M. Billig (Eds.), *After postmodernism: Reconstructing ideology critique* (pp. 58–78). London, UK: Sage.

Gergen, K. J. (1994b). *Reality and relationships: Soundings in social construction*. Cambridge: Harvard University Press.

Gergen, K. J. (1995a). Metaphor and monophony in the twentieth century psychology of emotions. *History of the Human Sciences, 8*, 1–23.

Gergen, K. J. (1995b). Social construction and the educational process. In L. Steffe & J. Gale (Eds.), *Constructivism in education* (pp. 17–39). Hillsdale, NJ: Erlbaum.

Gergen, K. J. (1996). Technology and the self: From the essential to the sublime. In D. Grodin & T. Lindlof (Eds.), *Constructing the self in a mediated age* (pp. 127–140). Beverly Hills, CA: Sage.

Gergen, K. J., Gulerce, A., Lock, A., & Misra, G. (1996). Psychological science in cultural context. *American Psychologist, 51*, 496–503.

Gergen, K. J., Hoffman, L., & Anderson, H. (1997). Is diagnosis a disaster? A constructionist dialogue. In F. Kaslow (Ed.), *Handbook of relational diagnosis*.

Gergen, K. J. (1999a). *An invitation to social construction*. London, UK: Sage.

Gergen, K. J. (1999b). The self: Death by technology. *The Hedgehog Review, 1*, 25–34.

Gergen, K. J. (2001). Psychology in a postmodern context. *American Psychologist, 56*, 803–813.

Gergen, K. J. (2002). Cell phone technology and the realm of absent presence. In D. Katz & M. Aakhus (Eds.), *Perpetual contact* (pp. 227–241). New York, NY: Cambridge University Press.

Gergen, K. J. (2003). A civil world beyond individual and community. In E. Brix & P. Brix (Eds.), *Zivilgesellschaft zwischen Liberalismus und Kommunitarismus* (pp. 191–204). Vienna, Austria: Passagen Verlag.

Gergen, K. J. (2005). *Therapeutic realities, collaboration, oppression and relational flow.* Chagrin Falls, OH: Taos Institute Publications.
Gergen, K. J. (2007a). *Construccionismo social: Aportes para el debate y la práctica* (M. Estrada Mesa & S. Diazgranados Ferrans, Eds.). Bogota, Colombia: Universidad de los Andes.
Gergen, K. J. (2007b). Relativism, religion, and relational being. *Common Knowledge*, special issue, A dictatorship of relativism, *13*(2–3), 362–378.
Gergen, K. J. (2007c). Writing as relationship in academic culture. In M. Zachry & C. Thralls (Eds.), *Communicative practices in workplaces and the professions* (pp. 95–102). Amityville, NY: Baywood.
Gergen, K. J. (2008). Mobile communication and the transformation of the democratic process. In J. E. Katz (Ed.), *Handbook of mobile communication studies* (pp. 297–310). Cambridge, MA: MIT Press.
Gergen, K. J. (2009). *Relational being, beyond self and community.* New York, NY: Oxford University Press.
Gergen, K. J. (2010a). The acculturated brain. *Theory and Psychology*, *20*, 795–816.
Gergen, K. J. (2010b). Co-constitution, causality, and confluence: Organizing in a world without entities. In T. Hernes & S. Maitlis (Eds.), *Process, sensemaking, and organizing* (pp. 1–26). Oxford, UK: Oxford University Press.
Gergen, K. J. (2011a). From moral autonomy to relational responsibility. *Zygon, Journal of Religion and Science*, *46*, 204–223.
Gergen, K. J. (2011b). From psychodynamics to multi-being. *Dynamic Psychiatry*, *3*, 251–252 & *4*, 93–119.
Gergen, K. J. (2011c). The social dimension of social psychology. In A. Kruglanski & W. Stroebe (Eds.), *Handbook of the history of social psychology* (pp. 137–158). New York, NY: Psychology Press.
Gergen, K. J. (2014). Pursuing excellence in qualitative inquiry. *Qualitative Psychology.*, *1*, 49–60.
Gergen, K. J. (2015a). From mirroring to world-making: Research as future forming. *Journal for the Theory of Social Behaviour*, *45*, 287–310.
Gergen, K. J. (2015b). Playland: Technology, self and social transformation. In J. De Mul, V. Frissen, M. deLange, & J. Raessens (Eds.), *Playful identities: The ludification of digital media cultures* (pp. 55–73). Amsterdam, The Netherlands: Amsterdam University Press.
Gergen, K. J. (2015c). Relational ethics in therapeutic practice. *Australian and New Zealand Journal of Family Therapy*, *36*, 409–418.

Gergen, K. J. (2015d). Toward a relational humanism. *Journal of Humanistic* Counseling, *54*, 149–165.

Gergen, K.J. (2015e). *An invitation to social construction*. 3rd. ed. London: Sage.

Gergen, K. J., & Basseches, M. (2013). Practices of psychological inquiry: The global challenge. In E. Mustakova-Possardt, M. Lyubansky, M. Basseches, & J. Oxenberg (Eds.), *Toward a socially responsible psychology for a global era* (pp. 47–64). New York, NY: Springer.

Gergen, K. J., & Davis, K. E. (Eds.). (1985). *The social construction of the person*. New York, NY: Springer-Verlag.

Gergen, K. J., & Gergen, M. (1981). *Social psychology*. New York, NY: Harcourt, Brace and Jovanovich.

Gergen, K. J., & Gergen, M. (1983). Narratives of the self. In K. E. Scheibe & T. R. Sarbin (Eds.), *Studies in social identity* (pp. 254–273). New York, NY: Praeger.

Gergen, K. J., & Gergen, M. (Eds.). (1984). *Historical social psychology*. Hillsdale, NJ: Erlbaum.

Gergen, K. J., & Gergen, M. (2010). Positive aging: Resilience and reconstruction. In P. Fry & C. Keyes (Eds.), *New frontiers in resilient aging: Life-strengths and well-being in later life* (pp. 340–356). New York, NY: Cambridge University Press.

Gergen, K. J., & Gergen, M. (2012). *Playing with purpose: Adventures in performative social science*. Walnut Creek, CA: Left Coast Press.

Gergen, K. J., & Marlowe, D. (Eds.). (1979). *Personality and social behavior*. Reading, MA: Addison-Wesley.

Gergen, K. J., & Ness, O. (2016). Therapeutic practice as social construction. In M. O'Reilly & J. N. Lester (Eds.), *The Palgrave handbook of adult mental health* (pp. 502–519). Basingstoke, UK: Palgrave Macmillan.

Gergen, K. J., & Shotter, J. (Eds.). (1989). *Texts of identity*. London, UK: Sage.

Gergen, K. J., & Thatchenkery, T. (1996). Organizational science in a postmodern context. *Journal of Applied Behavioral Science*, *32*, 356–377.

Gergen, K. J., & Whitney, D. (1996). Technologies of representation in the global corporation: Power and polyphony. In D. M. Boje, R. P. Gephart, & T. J. Thatchenkery (Eds.), *Postmodern management and organization theory* (pp. 331–357). Thousand Oaks, CA: Sage.

Gergen, K. J., & Zielke, B. (2006). Theory in action. *Theory and Psychology*, *16*(3), 299–309.

Gergen, K. J., Ellsworth, P., Maslach, C., & Seipel, M. (1975). Obligation, donor resources and reactions to receiving aid in three nations. *Journal of Personality and Social Psychology*, *31*, 390–400.

Gergen, K. J., Hepburn, A., & Comer, D. (Eds.). (1986). The hermeneutics of personality description. *Journal of Personality and Social Psychology, 6,* 1261–1270.

Gergen, K. J., Stam, H. J., & Rogers, T. B. (Eds.). (1987). *Metapsychology: The analysis of psychological theory.* New York, NY: Praeger, 1987.

Gergen, K. J., Gergen, M., & Barrett, F. (2004). Dialogue: Life and death of the organization. In D. Grant, C. Hardy, C. Oswick, & L. Putnam (Eds.), *Handbook of organizational discourse* (pp. 39–59). London, UK: Sage.

Gergen, K. J., Schrader, S., & Gergen, M. (Eds.) (2008). *Constructing worlds together: Interpersonal communication as relational process.* Boston, MA: Pearson.

Gergen, K. J., Dragonas, T., McNamee, S., & Tseliou, E. (Eds.) (2015a). *Education as social construction.* Chagrin Falls, OH: WorldShare Books.

Gergen, K. J., Josselson, R., & Freeman, M. (2015b). The promises of qualitative inquiry. *American Psychologist, 70,* 1–10.

Gergen, M., & Gergen, K. J. (Eds.) (2003). *Social construction: A reader.* London, UK: Sage.

Gordon, C., & Gergen, K. J. (Eds.) (1968). *The self in social interaction: Classic and current perspectives.* New York, NY: Wiley & Sons.

Graumann, C. F., & Gergen, K. J. (Eds.). (1996). *Psychological discourse in historical perspective.* New York, NY: Cambridge University Press, 1996.

Hersted, L., & Gergen, K. J. (2013). *Relational leading: Practices for dialogically based collaboration.* Chagrin Falls, OH: Taos Institute Publications.

Hosking, D. M., Dachler, H. P., & Gergen, K. J. (Eds.). (1995). *Management and organization: Relational alternatives to individualism.* Aldershot, UK: Avebury.

Jones, E. E., Davis, K. E., & Gergen, K. J. (1961). Role playing variations and their informational value for person perception. *Journal of Abnormal and Social Psychology, 53,* 302–310.

Kwee, M. G. T., Gergen, K. J., & Koshikawa, F. (Eds.). (2006). *Horizons in Buddhist psychology.* Chagrin Falls, OH: Taos Institute Publications.

McNamee, S., & Gergen, K. J. (Eds.) (1991). *Therapy as social construction.* London, UK: Sage.

McNamee, S., & Gergen, K. J. (1999). *Relational responsibility: Resources for sustainable dialogue.* Thousand Oaks, CA: Sage.

Semin, G., & Gergen, K. J. (Eds.). (1990). *Everyday understanding: Social and scientific implications.* London, UK: Sage, 1990.

Strickland, L. H., Aboud, F., & Gergen, K. J. (Eds.) (1977). *Social psychology in transition.* New York, NY: Plenum.

Sugiman, T., Gergen, K. J., Wagner, W., & Yamada, Y. (Eds.) (2008). *Meaning in action: Constructions, narratives, and representations*. New York, NY: Haworth.

Mark Freeman is Professor and Chair of the Department of Psychology and Distinguished Professor of Ethics and Society at the College of the Holy Cross in Worcester, Massachusetts. His writings include *Rewriting the Self: History, Memory, Narrative; Finding the Muse: A Sociopsychological Inquiry into the Conditions of Artistic Creativity; Hindsight: The Promise and Peril of Looking Backward; The Priority of the Other: Thinking and Living Beyond the Self;* and numerous articles on issues ranging from the psychology of memory and identity to the psychology of art and religion. Winner of the 2010 Theodore R. Sarbin Award in the Society for Theoretical and Philosophical Psychology, he is also a Fellow in the American Psychological Association and serves as editor for the Oxford University Press series "Explorations in Narrative Psychology."

The Incorrigible Science: A Conversation with James Lamiell

(Interviewed by Jack Martin, Simon Fraser University)

Historically, Lamiell's work links to ideas previously presented and debated by influential philosophers and psychologists of the past like William Stern, Wilhelm Wundt, and Wilhelm Windelband. Basic to Jim's work is the difficulty that most psychologists experience in distinguishing between statistical aggregates and the psychological lives of individual persons. Jim has been able to identify the many historical points at which the confusions that flow from this difficulty have been articulated and then ignored, or misunderstood, by historically significant research psychologists and most of the rest of us. In doing so, he's been able to describe the surprisingly consistent set of ploys by which criticisms of the prototypical practices of psychological inquiry have been deflected, skirted, and avoided without any really serious institutional attempt to respond to them and face them directly. And these, of

J. Lamiell (✉)
Georgetown university, Washington, D.C., USA
e-mail: lamiellj@georgetown.edu

H. Macdonald et al. (eds.), *Dialogues at the Edge of American Psychological Discourse*, Palgrave Studies in the Theory and History of Psychology, DOI 10.1057/978-1-137-59096-1_8

course, are the practices that continue to dominate the scientific arm of the discipline of psychology.

Throughout his career, Jim has attempted to reform psychologists' misunderstandings of statistical methods and misinterpretations of the data they yield and to encourage exploration of alternative quantitative and qualitative tools. After his graduate work at Kansas State University, Jim spent several years at the University of Illinois before settling at Georgetown University for the duration of his career in the academy. During his sabbatical years at Georgetown, Jim studied and worked in Germany, learning the German language so that he could read the original works of early German psychologists like Wilhelm Wundt and William Stern. As a three-time Fulbright Scholar he became an expert on the work of Stern and, as the Ernst Cassirer Visiting Professor at the University of Hamburg in 2004, he delivered a series of public lectures on the life and works of Stern. Jim's books (*Beyond Individual and Group Differences: Human Individuality, Scientific Psychology* (2003), *and William Stern's Critical Personalism and William Stern: A Brief Introduction to his Life and Works* (2010)) are essential reading for anyone seriously interested in the philosophy and history of psychology.

At the time of this interview, Jim had just completed a second term as Chair of the Department of Psychology at Georgetown University and was preparing to retire from formal academic life to continue his scholarly efforts with less disruption and even greater focus.

Jack Martin (JM): Jim, it's a great pleasure to have the chance to talk to you this morning about your work. Since the mid-1980s, you have pursued what I consider to be one of the most important and sustained critical challenges to psychological science. I want to spend most of our time during this conversation exploring your perspective on your work and the kinds of reaction it and you have received. As we go along, I hope to give you a chance to explain in some detail the nature of your concerns, your criticisms, and your pleas for change in certain basic assumptions that almost all psychologists tend to make, whether they are research psychologists or practitioners. In particular, I'd like you to describe your concerns about the relationship between the data typically yielded by psychological research and the statistical methods and interpretative lenses that have been used to think about and write about that body of research—research that is understood to support the scientist-practitioner model promoted by almost all major psychological associations and assumed by most psychologists, their clients, and the general public. All these people believe that psychological research supports the effectiveness of psychological interventions. However, your work challenges this basic assumption, and in doing so, challenges the ideas of psychology as a science and applied psychology as an applied science. In effect, you're asking psychologists to look very, very carefully at their science to determine what is reasonable to say and do based on that science and what is not. It's my great pleasure to talk to you today about all of this, Jim.

Jim Lamiell (JL): I'm really pleased to be here, and thankful to you all for this opportunity.

JM: You've obviously been at this for a considerable period of time.To begin, I think it would be useful for us to know more about just what drew you to this particular line of critique in the first place?

JL: Sure. As a graduate student, I pursued a course of training in the sub-discipline of personality psychology, understood then—and now—as a sub-discipline concerned primarily with the assessment and study of individual differences. As I familiarized myself with the literature of the field, it became pretty clear to me that in order to make one's way in the discipline, one really needed to gain as much skill as possible with statistical methods. So I spent a lot of time on that subject while in graduate school, and I have drawn extensively on what I learned then, though in ways that I did not anticipate at the time.

My first post-Ph.D. appointment was at the University of Illinois at Urbana-Champaign, where I joined the Psychology Department faculty in 1976. I was part of a team that offered a basic course in personality psychology, and my lectures were focused on the so-called "trait" approach to the study of personality. At that time, Walter Mischel's book, *Personality and Assessment,* which had been published in 1968 (Mischel, 1968), was very much at the center of discourse within the field. In that book Mischel reviewed the then extant research literature on the "trait" approach to the study of personality, and argued that if one looked at the empirical evidence that had accumulated to over the years up to then, the correlations between measures of individual differences in traits and measures of individual differences in the behaviors that should have been predictable on the basis of the trait measures rarely exceeded $r = .3$. This meant, of course, that the field's best trait measures were accounting for some nine percent (9%) of the variance in the criterion behaviors. That being the case, Mischel argued, there was good reason to doubt the scientific viability of the concept of personality traits, and to many in the field that was tantamount to questioning the existence of personality itself.

So I was presenting this material to undergraduate students at Illinois, trying to convey the notion that explaining a mere 9% of the between-person variance in criterion behaviors in terms of variables marking individual differences in personality traits was not very favorable to the theoretical assumption of appreciable consistency in the behavioral manifestation of those traits. It was for this reason I sought to make clear, that the scientific validity of the trait concept itself had been thrown into doubt.

Every time I presented this material, my students would plead with me to help them understand a bit more clearly just what the statistical index "percent variance accounted for" means with respect to the question of consistency in individuals' respective manifestations of trait-relevant criterion behaviors. "Does that index mean," I would often be asked, "that 9 percent of the people are consistent and 91 percent are not?"

"Well no," I had to say, "it doesn't exactly mean that."

"Well, does it mean that each person is consistent 9 percent of the time and not consistent 91 percent of the time?"

Of course I had to say "Well, no, it really doesn't mean that, either."

"Well, please, Professor, tell us! How are we to understand the meaning of the statistic "percent variance accounted for" with respect to the question at hand, namely, the consistency of individuals over time and across situations?"

JM: Because the students, of course, are thinking that psychology really is telling them something about individual persons.

JL: Exactly, and especially in a course in personality! But at this point in my lectures I always found myself in the situation of finally having to ask the students to simply take my word for it that this is the way the question of (in)consistency in the behavioral manifestation of personality traits is properly addressed from a scientific standpoint. I reassured them that once they had taken more coursework in statistical methods, they would see the sense in this. In the meantime, I would say, "please just take my word for it."

That ploy got me out of the lecture hall with my professorial dignity intact. But the exchanges always bothered me, and as I would return from the lecture hall to my office, I would regularly ask myself: "Why can't I answer their questions in the way in which they are put to me?" Why did I have to pull rank on them, as it were (although at that time I wasn't very much older than they were), and essentially say to them that their thinking was not yet sophisticated enough to formulate their own question properly? I always had to evade the question by reformulating it in such a way that the statistical index "percent variance accounted for" would be the answer!

Then one day, it struck me that I couldn't answer their question because with respect to individual level doings, indices of "percent variance accounted for" and the correlation coefficients on which those indices are based don't mean anything at all. As empirical grounds for knowledge claims, they are un-interpretable at the level of the individual.

That was truly a "Eureka!" moment for me, and it was then that I decided to write a journal-length article on this topic. I was intent on publishing my article in the *American Psychologist,* and this took two years to accomplish. Three versions were rejected before the fourth was finally accepted. Being, as I was, a young and untenured assistant professor at the time, one of my senior colleagues advised me to put my project on ice for the time being and to concentrate on other, more mainstream, work that could be published and earn me promotion and tenure. But I decided that that was something I could not do. So I continued to work on my article, and it finally appeared in the March, 1981 issue of the *American Psychologist* (Lamiell, 1981).

JM: And then you were off and running?

JL: And I was off and running.

JM: But most people weren't off and running with you?

JL: No.

JM: While you were talking, I was thinking about my own decision to go into psychology. Much of my undergraduate degree was devoted to the study of mathematics and physics. So it wasn't until graduate school that I looked at psychological research seriously for the first time and I remember thinking to myself, "You have to be kidding me. I mean, in physics the mathematics of calculus map onto the world, whereas the statistics of psychologists seem to retreat from the world."

JL: Quite so.

JM: I have had some personal experience of trying to convince psychologists of the limitations of their statistical methods. So I can imagine at least some of the resistance you must have experienced in your work. Can you tell us a few stories about that and how those reactions have affected you?

JL: Sure. I was actually pretty encouraged at the beginning. The *American Psychologist* article received a fair amount of attention.

> Most of it conveyed resistance to my ideas, but at the time, I was pretty happy because I thought, "Well, the dialogue has been started and this is what I wanted to happen." Since I was pretty sure that I was right about what I was saying, I was confident that, eventually, the mainstream would soon come around.

But that proved to be naïve. It didn't happen that way. After the initial flurry of dialogue, both at conferences and in print, the conversation seemed to just die out, and the field went on with business as usual. In short: there was criticism, I rejoined that criticism, and then the conversation just stopped.

JM: Nothing?

JL: Well, virtually nothing. In the ensuing years, part of the resistance that I have continued to encounter is what I would call "passive" resistance. It entails the continuation of traditional practices without addressing my critique or even acknowledging its existence. In confidence, some colleagues have actually said to me words to the effect that they know I am right but they are too committed to the traditional way of thinking to abandon it now.

Other pockets of resistance, which I would characterize as more active in that they entailed attempts to refute my arguments, led me at first to the view that my critics did not understand me, and that it was therefore incumbent upon me to recast my argument and try to state it more clearly. This I sought to do in many different publications (e.g., Lamiell, 1982, 1986a, 1986b, 1987). But I eventually came to see that a major reason that my critics did not understand me was that they did not understand themselves. What I mean by this is that, through statements made to me verbally and directed at me in print, I could see that, however sophisticated my critics may have been *technically*, they did not have a clear *conceptual* grasp of their own statistical methods.

I offer here just one example among many that I could cite. At a conference I attended some time in the 1980s, I listened to an invited talk given by a very prominent personality researcher of the time. In the first part of his talk, he took the occasion to chide another prominent researcher in the field—not present at the conference—who had drawn inferences about the doings of his individual research subjects on the basis of experimental treatment group means. The speaker was setting up his discussion of his own research findings, which would focus on correlations capturing, as they are suited to doing, systematic between-person variation around treatment group means.

At a little social gathering that took place after the presentation, I took the opportunity to chat briefly with the speaker. I told him that I appreciated his point about the inappropriateness of drawing inferences about individuals on the basis of group means. He seemed pleased. But then I said, "Does it trouble you at all that the correlation coefficients by which you are placing such great store in drawing inferences about your own individual research subjects are themselves group means?" From the look on my interlocutor's face, it seemed clear to me that this was a logical fact on the implications of which he had simply not reflected. He responded by saying, "Well, there are group means and then there are group means," and that was the end of the conversation.

Dating back at least to the time of Lee J. Cronbach's well-known article discussing the "two disciplines of scientific psychology (Cronbach 1957)," personality psychologists have thought of their correlational analyses as a vehicle for carrying them beyond knowledge of the effects of experimental treatments "on average," further into a more fine-grained knowledge of what was transpiring with the individuals who had been subjected to those treatments. What is overlooked in all of this is the fact that the index that is used to capture between-person variance around treatment group means, the Pearson product-moment correlation coefficient, *is itself a group mean*. As such, it is subject to the same interpretive constraints that limit the inferences we may make on the basis of experimental treatment group means. My interlocutor in the aforementioned exchange is far from the only personality investigator who has failed to appreciate the profound implications of this logical fact. Indeed, I think that those implications were lost on Cronbach himself. In any case, it is here where we find the crux of the problem.

To repeat for emphasis a point I made earlier: the lesson I took from this and many similar experiences is that a great many—perhaps even most—mainstream thinkers really do not have a firm conceptual grasp of their own statistical methods. This in turn made me realize that the problem to which I had pointed was—and is—much, much larger than I had judged it to be at the beginning, and I concluded that if there was to be any change in the views of mainstream thinkers, that change would be neither swift nor easy.

It was at about this time that I had a professional experience that turned my attention in the direction of more historical inquiry. In the spring of 1984, I was invited to participate in a symposium at the Second European Conference on Personality, which was scheduled to be held in the city of Bielefeld in what was then West Germany. The presentation that I made there was basically an elaboration of the arguments that I had been developing up to that point. After my talk, a German scholar named Lothar Laux (who has since become a close colleague and friend) asked me if I had ever heard of William Stern." I replied, "Oh, yes, he was the IQ guy." Laux rolled his eyes a little bit, but then patiently explained to me that Stern had done much more than that over the course of his career, and he, Laux, thought that, given what I had said in my presentation, I would find much of interest and value in Stern's work.

Words to the same effect were subsequently said to me by several other European scholars in attendance at the conference, and this finally persuaded me that I needed to look into the works of William Stern (1871–1938). My problem then was that most of Stern's works have never been published in English translation and I did not know German. Negotiating that hurdle required substantial time and effort, but by the time of my first sabbatical, which was spent in Heidelberg, Germany in 1990, I had progressed to the point where, slowly, and with a good dictionary next to me, I could begin to read Stern's writings. What an enlightenment this was for me! Stern, I learned, was actually the founder of differential psychology. In 1911, he published a textbook titled (in translation) *The Methodological Foundations of Differential Psychology* (Stern 1911), and as I read that work it became very clear to me that he had had a good grasp of the points that I was asserting some 70 years later! This both reinforced my resolve to push forward with my arguments and encouraged me to read further into Stern's works.

A key distinction drawn explicitly by Stern in his 1911 book is that between, on the one hand, knowledge about variables with respect to which individuals have been differentiated, and, on the other hand, knowledge about the individuals who have been differentiated in terms of those variables. The problem, as I was gradually coming to see it from an historical perspective, was that many of Stern's most influential contemporaries in the field were not clear on this distinction. Either they did not understand it, or, more ominously, they did understand it but then simply ignored it.

I found one very vivid example of this in a little book titled *Individuality,* published by Edward L. Thorndike (1874–1949) in 1911 (coincidentally the same year in which Stern's differential psychology book appeared). In the relevant passage, Thorndike is discussing the meaning of a correlation "between measures of two traits in a group of individuals," and he states that such a correlation indicates "the extent to which the amount of one trait possessed by an individual is bound up with the amount he possesses of some other trait" (Thorndike 1911, p. 21, emphasis added). So here, Thorndike is explicitly asserting that the aggregate-level correlation coefficient, which constitutes knowledge about the variables that have been correlated, also conveys knowledge about the individuals who have been differentiated from one another in terms of those variables.

A well-known work by another contemporary of Stern's, Hugo Münsterberg (1863–1915), who, by the way was a personal friend of Stern's, provides us with another example of this. In Münsterberg's *Psychology and Industrial Efficiency,* which was published in 1913 (Münsterberg 1913), there is a section where he refers to research findings establishing correlations between different aspects of attention. One particular passage states that

"persons who have a rather expansive span of attention for acoustical impressions have also a wide span for visual objects. Hence, the manifestation of one feature of attention by a person allows us to presuppose without further tests that certain other features may be expected in the particular individual" (Münsterberg 1913, pp. 135–136).

So here are two highly prominent differential psychologists, both of whom were contemporaries of Stern, who were claiming, in widely read and highly regarded works, that population-level statistical knowledge about variables with respect to which individuals had been differentiated is informative about some or another aspect of the psychological functioning of the individuals who have been differentiated in terms of those variables. These claims by Thorndike and by Münsterberg, respectively, are quite erroneous and they run directly counter to the distinction Stern had clearly drawn in his 1911 book. Nevertheless, it was the perspective shared by Thorndike and Münsterberg that came to dominate the field, and as that happened, Stern's distinction between the study of variables marking individual differences, on the one hand, and the study of individuals, on the other, was effectively obliterated. For if correlations between individual differences variables are, in fact, interpretable in the ways asserted by Thorndike and Münsterberg in the quotations just cited, then knowledge about the variables with respect to which individuals have been differentiated just *is*, at one and the same time, knowledge about the individuals who have been differentiated in terms of those variables. And that's what I mean by saying that the distinction that Stern drew was obliterated.

JM: It's completely conflated.

JL: Completely conflated. But that is the view that won out in the field. The further I have looked into the history of this development, the more I have come to the view that a major factor has been a long-standing lack of clarity about, or concern for, the distinction between frequentist and subjectivist thinking about statistical knowledge. Frequentist thinking recognizes that claims to probabilistic knowledge based on population-level statistical analyses are valid only for populations, and cannot validly be regarded as warrant for knowledge claims about individuals. Subjectivist thinking is reflected in statements of belief about the likelihood that thus-and-so is or will be found to be true of some individual, with the clear understanding that statements of belief about individuals are not and ought not be mistaken for claims to objective, empirical knowledge about those individuals.

One of the most surprising and discouraging passages I have ever encountered in my readings of publications relevant to these issues is one that I found in a 1989 book by the late Michael Cowles, titled *Statistics in Psychology: An Historical Perspective.* There is a section in his book where he is discussing the frequentist-subjectivist distinction, and the passage in his treatment of the matter that so struck me reads as follows:

> "[T]he fact that probability has to do both with frequencies and with degrees of belief is the . . . epistemological duality that . . . we [psychologists] blur as we compute our statistics and speak of the confidence we have in our results. The fact that the answer to the question, 'Who, in practice, cares?' is, 'Probably very few,' is based on an admittedly informal frequency analysis but it is one in which we can believe!" [Cowles 1989, p. 59]

So here we find a clear acknowledgement that psychologists have systematically blurred the frequentist-subjectivist distinction over the years, followed by the suggestion that since, historically, not many within the field have been concerned about this, there's no need for us to be concerned about it now.

I was *dumbfounded.* There most certainly *is* a need to deal with this blurring now, and the fact that it has been blithely indulged within the discipline by so many for so long has only made coming to terms with it that much more difficult.

The nature of this difficulty was clearly illustrated to me in an exchange I just had with Louis Sass, who happens to be here with us today. He began by proposing that we look at things from what he called "a common sense point of view." Suppose, he said, that an observer finds across numerous encounters with many different individuals over time that some trait T often, even if not always, goes with some other characteristic, C. So, in that observer's next encounter with an individual who displays trait T, the observer thinks, based on his/her experience, that it is likely that that same person will display characteristic C as well. Sass noted that this is just the sort of thinking practiced by the mainstream psychologists whom I am criticizing, and as if speaking for those countless mainstream thinkers, he asked me to explain further just what I find wrong with such thinking.

My response to this request runs as follows: In and of itself, there is nothing wrong with the sort of thinking described. It has been said that an observer's experiences with many people prior to encountering person P have inclined that observer *to think it likely* that person P will display some particular trait or characteristic, C. This is an example of subjectivist thinking. It is a statement about what the observer thinks about person P, and not a claim by the observer to have knowledge of an objective empirical fact about person P. The statement "observer thinks it likely" expresses a belief that the observer has about what might prove to be the case about person P.

So far, so good. The problem lies in the failure to remember that the probability that person P will display characteristic C is *always* one or zero—the characteristic either will or will not be displayed—and that this is true *no matter what* the frequency distribution is in the data that the observer has accumulated through prior experiences, *no matter what* that observer's subjective belief about person P happens to be, and regardless of whether that belief is based on those prior experiences or not. These are the crucial points that get lost when psychologists blur the distinction between frequentist and subjectivist understandings of probability, and, having done so, deliver themselves of proclamations such as "research establishes that the probability that person P will display characteristic C is thus-and-so." The research establishes nothing of the sort, and the present challenge for us as scientific psychologists is to insist on absolute clarity in such matters. *No more blurring allowed!*

Seeming still a bit puzzled, Sass asked me to consider another example with a slightly different twist. Suppose, he said, that a medical doctor had consistently been giving penicillin to patients with certain symptoms, and had found over time that 85 percent of those patients improved. Might not the doctor be well advised, Sass asked, to continue to administer penicillin to patients showing those same symptoms? Moreover, might not any given patient rather be given the penicillin than not?

Here again: by posing these questions to me, Sass was imploring me to explain further what I find problematic about such thinking, and here again I start by saying that, in and of itself, there is nothing wrong with such thinking at all. Yes, the doctor might, indeed, be well advised to continue giving his patients penicillin, believing on the basis of the already accumulated data that, going forward, 85 percent of the patients would continue to get better. But note that "85 percent patient improvement" is an empirical fact about a *population,* and might well serve actuarially as a basis for the prediction that in a replication study, a sample of patients drawn from the same population will, once again, manifest an 85 percent improvement rate. This is frequentist thinking *par excellence.*

Here again, however, there is nothing in knowledge about the percentage improvement rate within a population that would justify a claim by the doctor to know that *this* patient standing in the office today "has" an 85 percent chance of improving. The data in question entitle no such knowledge claim at all. *This* patient's chances of getting better are one or zero, and this is true *no matter what* the improvement rate within the population has thus far been found to be, and *no matter what* the doctor happens to believe in this or any other individual case.

Continuing with Sass's example, the truth of this latter point should also be made clear to the patient who expresses the preference to have rather than not have the penicillin. That is, the patient should be counseled to understand that it has by no means been scientifically established that *his/her* chances of getting better by taking the penicillin are 85 percent. They are one or zero, and if, knowing that, the patient still feels, subjectively, that taking the penicillin is better than not taking it, then so be it. But whatever choice the patient makes, the probability of improvement in his/her individual case will remain one or zero.

I'm also reminded here of a passage in the marvelous book by the historian Theodore Porter, titled *The Rise of Statistical Thinking: 1820–1900* (Porter, 1986). In that book, Porter writes of the views of the French physiologist Claude Bernard (1813–1878). Addressing himself *in 1865* to the very point under discussion here, Bernard wrote:

"Statistics can allow [the doctor] to tell [his/her] patient that, of every hundred such cases, eighty are cured . . . but that will scarcely move him. What he wants to know is whether he is numbered among those who are cured" (Bernard, 1865, quoted in translation from the French by Porter, 1986, p. 160). Please note again the date of this quotation: 1865!

Perhaps further clarity on this point can be achieved by considering the writing of the logician John Venn (1834–1923). In 1888, Venn, published his book *The Logic of Chance* (Venn 1888), and in that book explained that probabilistic knowledge claims cannot sensibly be divorced from the consideration of a *series* of cases. So when you say "85 percent of the patients given penicillin have been found to get better," that means "85 *out of a series of 100,*" or "850 *out of a series of 1,000,*" or what have you.This is the very essence of frequentist thinking. But as a claim to knowledge about some given individual, the statement "85 percent of the patients given penicillin have been found to get better" simply makes no sense. An individual case is not a series of cases. Moreover, and precisely because probabilitstic knowledge claims only have meaning in the consideration of a *series*, a statement of the sort "the probability is .85 that *this* person will get better" must likewise be regarded as either meaningless or false: the probability that, given penicillin, *this* person will get better is one or zero, and, again, this is true no matter what studies have revealed about the percentage of individuals within some indefinitely large group who get better.

I repeat for emphasis: blurring of the sort that Cowles identified as routine within psychology and then dismissed so cavalierly as unworthy of our concern is something that should no longer be tolerated in our science.

JM: All of this is interesting in part because Wundt and the other early experimentalists in psychology didn't use statistical analyses in the now commonly accepted ways, but nevertheless understood their work as a quest for knowledge that could be generalized in the sense of being common to all of those investigated. In other words, each person that was tested had to reveal the same response pattern or trajectory as that found in other tested individuals, right?

Jim Lamiell: Yes, if the phenomenon under investigation was to be regarded as generalizable.

JM: And so the early German experimental psychologists, were not confused about these kinds of things.

JL: No. Statistical analyses of variables defined for populations simply were not a part of the scientific program established by the early experimental psychologists. I'll come back to this point, but here I'd like to mention that another prominent person early in the twentieth century, and actually a devotee of Stern's work, was Gordon Allport. His is an interesting case because, although there are places in Allport's writings which suggest that he understood the inappropriateness of drawing inferences about individuals on the basis of statistical analyses of variables defined only for populations, there are other places where his writings are confused on this point. So even though he was a harsh critic of mainstream thinking in personality psychology, the record on these matters in Allport's case is decidedly mixed. Indeed, he eventually felt compelled to concede the mainstream argument that knowledge gained from the statistical analyses of variables defined for populations was to some extent informative about individual-level doings, and his concession on that point is what would finally force him to, as he himself put it, "cry uncle" and retire to his corner (Allport 1966).

JM: Didn't Allport understand personality research as providing a general backdrop against which an individual's personality might be considered? So when we're looking at Jack Martin's personality, we look at it in terms of the established background that the science of personality has established, the dimensionality, and so on?

JL: Right, that's exactly right. In his 1961 book (Allport 1961) explicitly stated that he was "not condemning the common trait approach" (p. 360), and he acknowledged that approach as being useful for identifying the "elements" of personality. But he went on to urge acceptance of the view that "personality exists only at a post-elementary state . . . when the common features of human nature have already interacted with one another and produced unique, self-continuing and evolving systems" (p. 161). So while Allport insisted that individual differences research could not tell the *whole* story about individual personalities, and so required supplementation through inquiry of a more idiographic nature, he did concede that individual differences could tell *part* of the story. That was his downfall. The argument that he needed to make, and that I am making, is that, no, individual differences research does *not* tell even part of the story. It tells an altogether separate story. Individual differences, if by that we mean between-person differences, are not definable at the level of the individual, and that is why the systematic study of such differences quite literally provides knowledge of *no one.*

JM: Yeah, the word individual, used as an adjective, gives a false impression.

JL: It does, indeed. You see, in order to talk coherently about an individual *difference*, one needs to be regarding, minimally, two people, and the stated difference between them cannot properly be attributed to either one of them individually (Lamiell 1990a, 1990b, 1997). That is what I mean when I say that individual differences do not exist at the level of the individual, and that is why individual differences research does not tell even a part of the story about individuals.

Now, if I might briefly return to Wundt. He and the other original experimental psychologists were doing what they called individual psychology. And when Stern in his 1900 book on differential psychology (Stern 1900) was casting about for a term to label the new sub-discipline he was proposing, he noted that the term "individual psychology" was already in use. Clearly, he had in mind Wundt and the experimentalists.

Experimental psychology to the original experimentalists was what we would call nowadays an "N = 1" affair. Laboratory investigations were carried out on individual subjects, one at a time, and the findings of those investigations were defined for and linked to specific individuals. Moreover, all of this was done in the quest for knowledge of the general laws presumed to govern mental processes. This is most puzzling to today's thinkers, who are inclined to wonder how general laws can possibly be sought through the study of individual cases. Puzzlement on this point dissolves once one appreciates what the early experimentalists meant by "general."

At its beginning in Germany, "individual" psychology and "general" psychology were one and the same project, a project commonly referred to as *die allgemeine Psychologie*. The key word here is *allgemein*, which is the German word for "general" (in the non-military sense under consideration here). That word evolved in the German language as a contracted form of the expression *allen gemein*, which means "common to all." If you're interested in discovering lawful regularities that are *allen gemein*, common to all, and if the entities of interest are individual organisms, as most certainly was the case for the early experimentalists, how would you go about the search for such laws *except* by studying individual cases? The work by Ebbinghaus (who was, incidentally, one of Stern's mentors during Stern's doctoral studies in Berlin) is a paradigmatic case. To establish that the "forgetting curve" Ebbinghaus famously discovered, through tests that he conducted on himself, represents a general law, one would have to show in studies of many individuals *studied one at a time* that, in case after individual case, the data arrayed in accordance with that curve. Only in this way could evidence accrue that the phenomenon is *allen gemein,* or *common to all* of the individuals investigated.

But now as statistical methods made their way into the field of psychology, the meaning of "general" changed. Talk of what is "generally" true ceased to be talk about what is common to all individuals, and became, instead, talk of what is "true on average" within some specified population. Quite obviously, these two meanings of "general" are radically distinct.

JM: Yes, and you can see that this is so if you do a scatter plot on the data from any psychological study of treatment effects. In these studies, data from different individuals are all over the place, right? The individual scores are all around the average. No wonder correlation coefficients account for such a minimal amount of overall variation.

JL: That's often true, but in the context of the present argument, it wouldn't matter if they accounted for much larger proportions of the variance. The treatment group means are still defined only for populations. Moreover, the correlations computed to capture variance around those treatment group means, being themselves averages, are, likewise, still defined only for populations, and hence still subject to the same interpretive constraints that apply to all averages. The only exceptions here are group means accompanied by non-zero variances or correlations that are perfect. Those are the only instances in which what is true on average is also true in general, and, of course, such cases never occur in actual research (cf. Lamiell 1998).

JM: Okay, fair enough. Good.

JL: Now, the thing of it is, and what must be mentioned here as well, since you brought up Wundt, is that already by the time that this shift in the meaning of "general" from "common to all" to "true on average" was taking place, proceedings leading to the divorce of psychology from philosophy were likewise already underway. This is important because philosophy is a discipline in which close *conceptual* analysis is prized, and the problem with which we are concerned here is a conceptual problem, not an empirical one. With this in mind, let us consider briefly here what Wundt had to say regarding this impending divorce. In the foreword to his 1913 essay published under the title (in translation) *Psychology's Struggle for Existence* (Wundt 2013), Wundt wrote:

"In the opinion of some, philosophy and psychology should divorce from each other.... [I]f this [divorce comes to pass], philosophy will lose more than it will win. But psychology will be damaged the most. Hence, the argument over the question of whether or not psychology is or is not a philosophical science is for psychology a struggle for its very existence."

Wundt was saying to the psychologists of his day, "A divorce of psychology from philosophy is a bad idea, and it will quite possibly lead ultimately to psychology's demise." Wundt, himself a philosopher, well understood that one of the most important functions of philosophy is its quest for conceptual clarity. What do we mean when we say X? He made clear his great doubt that empirically-oriented psychologists concentrating on the design and conduct of experiments, the gathering and analysis of data, and so on would be able, on their own, to maintain an appreciation for and the skills needed to deal with conceptual questions. On this development, he believed, psychology would eventually founder.

My view is that this is what, in fact, has happened in psychology. Allow me to share with you an anecdote from my own professional life, one directly relevant to this conversation. Within a month or so after my *American Psychologist* article appeared in March of 1981, I encountered in the coffee room a senior and very influential person in my department at the University of Illinois (1981). He congratulated me on the publication of my *American Psychologist* article, and I thanked him. As he turned to leave the room, he tossed back over his shoulder, "But it's merely theoretical." It was his use of the term "merely" in that instance that made me realize that I was not going to be many years at Illinois. I didn't know Wundt's piece at the time, but now I view that experience as one very vivid example of his prescience. I do think that if we look at the institution of psychology—our institutions—what we see is a discipline in which theoretical and philosophical discourse has been greatly devalued.

Consider the small size of Division 24 in APA. Or consider what is essential nowadays for having a successful career in mainstream psychology: conducting empirical investigations, experimental and otherwise, according to the received methodological canon, and securing the grant money to support those activities. So, yes, I now view Wundt's 1913 piece as extraordinarily prescient in this regard.

JM: You know, the mere fact that these critiques have been around for so long is astounding. For mainstream psychology, it's been business as usual in the face of these criticisms—criticisms of the most basic methods of the so-called scientific approach in psychology. Of course, we all know the historical stories of how psychology adopted and adapted scientific methodology, or what it took to be scientific methodology, as a basis for making its claim that it was a science. In many ways, what you're talking about is an instance of the ways in which methodology often has defined psychological investigation in the absence of any clear conceptualization of the subject matter of psychology. It's as if the methodological tools are determining the entire enterprise.

JL: Very much so, very much so. I'm certainly not the first or only one to talk about the way in which—and this is on your point, I think—the way in which psychology has made a metaphysics of its methods. The point of positivism, a philosophy of science that achieved widespread favor within mainstream psychology, was to rid psychology of metaphysics, but metaphysics has come back in the form of method.

A chapter that I've written on this very topic has just recently appeared in print (Lamiell 2016; see also Lamiell 2015). In that work, I write about the need for clarity with respect to what we as psychologists mean when we speak about "effects." We speak of empirical investigations in which we carry out statistical analyses on our data in search of treatment "effects," or even "effects" due to such non-experimental factors as sex, race, age, etc. But when in my graduate seminar I pose the question, as I have done over the years: What do you mean by "effects?" the first response I have regularly gotten has come in the form of looks of incredulity that seem to say "What kind of a question is that?" Pressed to answer verbally, the students would regularly make statements such as: "We do our statistical analysis and we speak of an "effect" if there is a statistically significant relationship between the variables we are examining." And I respond: "Well, OK, but who or what are the entities that are being affected by these "effects"? Where, exactly, are these effects being realized? And: What, exactly, is it about your statistical analysis that enables you to see the "effects" that you claim to see?"

Absent answers to these questions, the expression "effect" has no meaning apart from the statistical analysis that putatively reveals it. There isn't anything more substantive to be said about it. This is what it means to make a metaphysics of method. The methods textbooks teach the students the calculations necessary to search for these "effects," but I have yet to find a textbook—and I've looked pretty hard—that discusses the psychological meaning of these "effects," i.e., that addresses the questions I posed above.

JM: When I teach your work in an advanced seminar on theory and history of psychology, populated by both advanced undergraduate and graduate students, it takes a while for the penny to drop. But when the students begin to understand that group differences don't have anything to do with understanding particular individuals, they become really concerned. And then they want an alternative. "Well okay, so this doesn't work. Well, what are we supposed to do? We have to have a scientific basis for our actions. But if Lamiell's right, then what's the alternative?" They want an alternative so that they can continue to conduct their business more or less as usual. You must have faced this many, many times?

JL: I have, and my response to my interlocutors has several facets. First, I say, if you recognize the validity of my critique, if it's persuasive to you, and you nevertheless continue using the methods that you're using because I haven't equipped you with a new hammer, you're part of the problem. That may seem harsh, but the point is that if these traditional practices, these paradigmatic practices, have been successfully discredited, they don't somehow become okay just in case a method adequate for the purposes of psychological inquiry is not immediately at hand. It seems to me that it is incumbent upon everyone in the field to join in the search for or creation of adequate methods. I don't see any reason why the task of doing that has to fall exclusively to the critics of conventional practices. That would be point number one.

A second point I would emphasize in this context is that my critique is predicated on the assumption that what we are after as psychologists is knowledge about individuals. This was certainly the case at the beginning of experimental psychology in the Wundtian laboratories, as I noted earlier. And I have yet to find any evidence that the agenda was changed. If that happened somewhere along the line, I didn't get the memo, so to speak.

Now, I have encountered colleagues who have responded to my critique by saying, "Well, you know, I'm not really interested in individuals anyway, but rather in finding out things that are true in general." After advising them that what they really must mean, and therefore what they should say, is that they are after knowledge of what is true *in the aggregate*, and hence for populations, I acknowledge that their research methods are formally suited to that goal, and encourage them to carry right on.

But before they go merrily on their way, I hasten to add that they have now defined their scientific objectives as those of a species of demography, which just is the study of populations. Once that is acknowledged and stated explicitly, then it can at least be said that their knowledge goals and their research methods are properly aligned.

JM: So if you're a state administrator in charge of mental health for the state of Illinois, you might be able to use these data for broad public planning purposes?

JL: Exactly so.

JM: As you would use other demographic kinds of data?

JL: Right.

JM: But if you're looking at what is going to be most useful for a particular individual, then this information wouldn't be of use to you at all?

JL: It would not be; that's correct. Demography is not psychology, and it is not a substitute for psychology. So for those whose knowledge objectives really are psychological and not demographic in nature, the challenge of abandoning traditional methods in favor of other methods actually suited to the task remains.

JM: And in fact, forcing people to use scientifically based intervention practices, empirically established treatments and so on, is a current trend into which all these misunderstandings flow. So the poor clinician who's trying to do the best for her client is having to pull on and rationalize all her efforts within a set of scientific understandings which are totally and absolutely irrelevant to her particular endeavor?

JL: I would certainly agree with that. Therein lurks an ethical question. I think that there is a facet of this entire argument that is not only epistemological. It's not *just* a matter of saying, "Look, we've got to get things conceptually right here." On the contrary, if we are put in situations institutionally to bring our practices into line with extant empirical evidence, and if we're trying to work with individuals but the evidence doesn't tell us about individuals, then we're really being forced into treating our patients, our clients, whatever the case may be, in ways that our science most certainly does *not* warrant and therefore ought not mandate. Moreover, and as has been suggested by our colleague Jeff Sugarman, who is present here today, these considerations are by no means relevant only within psychology. On the contrary, and has been further underscored by another colleague present here today, Todd Rose, they are increasingly becoming points of emphasis in medicine, in education, and in the military.

Statistical methods of the sort that have long dominated psychological research require that persons be regarded as instances of the categories used to define the variables under investigation. This entails a commitment to the view that every instance of a given category is completely interchangeable for any other instance of the same category. In this way, persons are made into things. Moreover, and just because the discourse that we use as putatively authoritative scientists gets picked up in the society around us, lay persons come to think of themselves as instances of categories. Thus, with our scientific assistance, lay persons come to see their very identities as persons in terms of the categories that they see themselves as instantiating, and, in effect, help make themselves into things. These considerations lead me to believe that if psychology can shed the blinders that its statistical concepts and methods have long since become, the possibilities for conceptualizing persons as other than things will be appreciably enhanced, redounding to the good of society as a whole.

In several of his writings during the first third of the twentieth century, Stern made known his concerns along these lines. For example, in some writings addressed to the practices of psycho-technicians—the aforementioned Hugo Münsterberg being one of the most prominent among them—Stern repeatedly voiced his concern that they were not only interceding in people's lives but actually interfering in their lives. This is where, for him, epistemological concerns shaded over into ethical concerns (e.g., Stern 1921, 1929, 1933; cf. Lamiel 2010).

If I might return now to a point I made earlier about psychology having become a species of demography, I'd like to mention in that connection the perspective on social science that was urged by the Belgian polymath Adolphe Quetelet (1796–1874). In the late 1700s and on into the 1800s, Quetelet was arguing for what he called a "social physics," driven by population-level statistical analyses. The key concept in Quetelet's thinking was *l'homme moyen,* the average man, and he understood quite clearly the difference between demography and psychology. Indeed, Quetelet was of the view that a scientific psychology was effectively impossible because individual behavior is too arbitrary and unpredictable to be grasped scientifically (Porter 1986). In the realm of human affairs, we do not find enough order to start talking scientifically until we go to the aggregate level.

In an 1835 publication, Quetelet addressed himself to this point as follows:

"If one seeks to establish in some way the basis of a social physics, it is he, *l'homme moyen,* whom one should consider without disturbing oneself with particular cases or anomalies and without studying whether some given individual can undergo a greater or lesser development in one of his faculties." [Quetelet, 1835, quoted in translation from the French by Porter, 1986, pp. 52–53]

So if psychologists claim that they are not really interested in individuals but are seeking knowledge that is more general—I note here again that psychologists regularly use the term "general" when, for reasons discussed earlier, they should be using the term "aggregate"—they're effectively following in the footsteps of Quetelet. As stated earlier, that is fine, but what such psychologists must finally come to grasp, as Quetelet clearly did, is that psychology is not demography. On the contrary,

Quetelet clearly understood the truth later expressed in the claim in 1867 by the German philosopher and mathematician Moritz Drobisch that

"It is only through a great failure of understanding [that] the mathematical fiction of an average man ... [can] be elaborated as if all individuals ... possess a real part of whatever obtains for this average person." [quoted in translation from the German by Porter, 1986, p. 171]

In short, an important part of my answer to interlocutors on the question: "Okay, Lamiell, what are the alternatives?" is itself a question: Do you want to do psychology or do you prefer to continue doing demography?

Another part of my answer to that question entails pointing to qualitative methods. Certainly, they are an avenue along which we can advance our psychological understanding of the doings of individuals, and Stern, too, even as the father of differential psychology, viewed qualitative methods as utterly indispensable for a successful psychology (cf. Lamiell 2003). So as quantitatively oriented as he was in some of his work, and as quantitatively sophisticated as he was, he was not blind to importance of qualitative work. Such considerations were part of Gordon Allport's arguments as well. Unfortunately, argumentation along this line has never set well within the mainstream, and I realize as I look out at the audience for this interview that, to a substantial degree, I'm preaching to the choir today. Many here present are advocates of and strong spokespersons for qualitative methods.

I am well aware that part of the opposition to calls for greater use of qualitative methods is that reliance on such methods compromises psychology's status as a science. Well, I think there's important critical work to be done on this argument as well. The term "science" has been co-opted by mainstream thinkers, so that now, the expression "scientific psychology" has come to refer strictly to psychology *on the model of natural science*. But not all of science is natural science, and at the time at which experimental psychology was founded, a much broader view existed.

In German, the word for science is *Wissenschaft*, and so to do science was to "*schaffen*," which means to do or to make, *Wissen*, or knowledge.

And just how one would go about the "*schaffen-ing of Wissen*"—i.e., what methods would be used to do this—was an open question. Some methods, chiefly experimentation and mathematization, were well suited to the *natural* sciences, *die Naturwissenschaften,* but other methods, chiefly qualitative in nature, were better suited to the *human* sciences, *die Geisteswissenschaften.* Unfortunately, the view ascended over the years that genuinely scientific knowledge in psychology could be secured only by proceeding on the model of the *natural* sciences, and this is a stance that needs to be challenged anew.

Now I know that there are others within our field who have mounted versions of this challenge in the past and have now come to wonder whether or not it is even necessary to couch some of their work under the umbrella of "science" at all. Mark Freeman, another colleague who is with us here today, has been involved in this effort, and in conversation with me has stated quite clearly his appreciation of the point that *Wissenschaft* as the making of knowledge must not only mean quantitative, experimental investigation, but also includes such disciplines as history and anthropology, and could even be taken to include the humanities generally. So Freeman has been led to ask himself, and, by extension us, such questions as: What advantage do we gain, other than, perhaps, linguistic credibility, by continuing to say "let's do science"? Why not just say we're doing something else? We're looking at individuals and we're trying to understand them!

Freeman's point is a good one, and merits our serious consideration. My own position, for now at least, is that there is still merit in trying to open up space for trying to understand the term "science" more broadly, and, indeed, in a way more faithful to nineteenth century German thought. Some people will say that if you're doing person-centered inquiry that relies on qualitative methods, it's not science. For them, that's the end of the discussion. So in order to get an audience there, it seems to me that it requires an expansion of the understanding of "science." It must be made clear that we are not necessarily abandoning science, but resurrecting an earlier understanding according to which the human sciences, *die Geisteswissenschaften,* were and are, indeed, disciplines that entail the "*schaffen*-ing of *Wissen,*" the making of knowledge. Mainstream thinkers have effectively arrogated the term "science" to suit there own narrow views, and I want to call them on that while arguing for an understanding of the term more in line with earlier thinking.

I do not want to suggest here that I think that Freeman has altogether abandoned the struggle. On the contrary, it seems clear to me that he still advocates a much broader use of the term "science," and I totally agree. But I also wish to emphasize that gaining such breadth is not a matter of endowing the term "science" with a meaning it never before had, but rather reclaiming for the term "science" a meaning it once *did* have but lost, at least within mainstream psychology.

In Wundt's time, philosophy was a science, and in his 1894 work, *Geschichte und Naturwissenschaft,* or *History and Natural Science,* Windelband distinguished the *empirical* sciences, which included *both* the natural sciences *and* the human sciences, from the *rational* sciences, disciplines having no specific empirical content, i.e., logic and mathematics. In this light, it actually angers me that mainstream psychologists have, as I said, co-opted the term "science" in the way I've described, making natural science the only acceptable model for anything deserving the brand "scientific." Mind you, I, myself, don't feel particularly needful of being regarded as a scientist, and perhaps, as Freeman has suggested in conversation with me, perhaps there is interesting territory to explore beyond the boundaries of science. But I'm still committed to the effort to re-claim for work of the sort we find needed the recognition that it is a form of "*schaffen*-ing" *Wissen,* i.e., making knowledge, and that is what science is. Taken to refer only to *natural* science, the idea of a "scientific psychology" becomes much more constrained than it could be, once was, and should be again.

All of that said, yet a further point to be made in this connection is that I have never been, and am not now, categorically antagonistic in principle to any and all uses of quantitative methods in psychological research. I think that there is a place for quantitative research in psychology. I think again, for example, of the early experimental psychologists. Fechnerian psychophysics had a high profile in Wundt's laboratory. Or consider again Ebbinghaus's experiments on memory: in his work one finds a great many equations, some of which, by the way, are statistical. But Ebbinghaus was using statistical calculations as a way of estimating measurement error, and all of the data that he was analyzing statistically came from a single subject—Ebbinghaus himself. So the problem is not with the use, *per se,* of quantitative methods, including statistical calculations. The problem is the reliance on the statistical analysis of population parameters as the basis for claims to knowledge about the causes or predictors of individual-level doings.

Peter Molenaar, at Penn State, is one researcher whose work exemplifies the use of very sophisticated quantitative methods in the study of individuals. Todd Rose, of Harvard is also making important contributions in this direction.

I myself did some highly quantitative work back in the 1980s, looking at the psychology of subjective personality judgments (see, e.g., Lamiell and Durbeck 1987; Lamiell et al., 1983). I'm not going to discuss that work in any detail here, but it is part of the published record, and does serve to illustrate one way in which theoretically significant questions can be fruitfully addressed quantitatively at the level of the individual.

I wish also to mention in this connection the work of James Grice (e.g., Grice 1911). He is a member of the Psychology Department faculty at Oklahoma State University, and he's developing what I consider to be some very useful tools for investigating individual level phenomena in a quantitative way.

So, to the student who says, "What are we supposed to do instead?" I want to say, "Listen, if you look a little bit, there are some real possibilities out there. None of them is the last word, but it's not as if no progress at all has been made on this issue."

JM: One of the things I often say to students when they ask me that question, or at least to help them understand what's at stake is to use an example from baseball. If you wanted, for example, to calculate the likelihood that a particular player in a certain game situation is going to get a hit, what would you do? Would you look at all the performance data of that individual over his career? Or would you categorize him as being Hispanic, left-handed, over six feet tall, et cetera—assuming that there were the equivalents of psychological data available on left-handers versus right-handers, Hispanics versus non-Hispanics, and so on? Well, quite clearly, any sane person would look at the performance across time of that particular individual. But that's not what psychology does. And we ought to realize that, right?

JL: Absolutely. It's not—it's not an anti-quantitative methods rant by any stretch of the imagination, and it doesn't have to be.

JM: Jim, I want to ask you one more question.Given all that you have said, I'm pretty pessimistic about the possibility that anything will change. In psychology, we have what seems to be nothing short of an almost purposeful obfuscation of the kinds of distinction and clarification you are insisting on. I'm going that far because I know you won't want to. Do you think there is any hope for this discipline—for psychology?

JL: In its super-condensed form, my answer to your question is: short term, little hope; long term, maybe a glimmer of hope. Let me elaborate just a little bit.

My whole effort in going down this road, dating back to those classroom experiences at Illinois about which I spoke at the beginning of this interview, has been aimed at *getting things right*. Alas, my experiences over these many years have led me to believe that that objective is not widely shared within the field, at least not nearly as widely as I think it needs to be shared. I see that both in graduate students whom I encounter, and in colleagues both within and outside of my own department. It appears to me that the predominant objective is *to have a career*, and the thinking seems to be that if the pursuit of that objective obstructs efforts to get things right, that's just the way it will have to be. I think that the blurring of the frequentist-subjectivist distinction that we discussed earlier is a very clear evidence of this thoroughly deplorable ethos. I don't see this state of affairs changing any time soon, and that is the primary basis for my short-term pessimism.

Viewing matters more long term, I do see at least some encouraging signs. One of them I have found in the establishment of a new journal called *The Journal for Person Oriented Research*. It is published in Sweden, and is linked with an organization called the Society for Person Oriented Research. The mission statement of this new journal includes the following passage:

"Person oriented research refers to theoretical methodological and empirical research that is guided by a research paradigm in which the individual is at focus and seen as a functioning totality. This paradigm implies that theories and findings should be interpretable at the level of the individual and that patterns of individuals' characteristics are of key interest. Hence, a standard variable oriented approach with the variable as the basic conceptual and analytic unit, and analyzing data using group statistics, for example, correlational analysis, falls normally outside the journal's scope."

So, as you can tell, I regard the appearance of this journal as a very welcome development. The mission statement does go on to say that the journal will not be receptive to research relying on qualitative methods, which –

JM: . . . is unfortunate.

JL: Yes, I think it is unfortunate, but again thanks to you, Jack, as well as Mark Freeman, Jeff Sugarman, and several others, some of who are here at this session, there are other efforts in this direction. So, thinking longer term, developments like this do give me some cause for optimism.

In the spirit of Wundt's 1913 essay discussed earlier, I think that one thing that's going to have to happen in psychology is that, discipline-wide, we are going to have to prioritize the objective of getting things right. We must bring our disciplinary house into conceptual order, and that means that within the departments in which we variously work, space must be made, both literally and figuratively, for scholars—both young ones and those more seasoned—who will tend to the conceptual work that is so vital for the intellectual health of the field.

All of that said, I want to thank you very much, Jack, for giving me this opportunity to talk about these ideas. I also thank Heather Macdonald for her important role in organizing this, and I thank you all for coming and being a part of this. I am honored and very grateful.

Bibliography

Allport, G. W. (1966). Traits revisited. *American Psychologist, 21*, 1–10.

Cowles, M. (1989). *Statistics in psychology: An historical perspective.* Hillsdale, NJ: Lawrence Erlbaum Associates.

Cronbach, L. J. (1957). The two disciplines of scientific psychology. *American Psychologist, 12*, 671–684.

Grice, J. W. (1911). *Observation oriented modeling: Analysis of cause in the behavioral sciences.* New York: Elsevier.

Lamiell, J. T. (1981). Toward an idiothetic psychology of personality. *American Psychologist, 36*, 276–289.

Lamiell, J. T. (1982). The case for an idiothetic psychology of personality: A conceptual and empirical foundation. In B. A. Maher & W. B. Maher (Eds.), *Progress in experimental personality research* (Vol. 11, pp. 1–84). New York, NY: Academic Press.

Lamiell, J. T. (1986a). Epistemological tenets of an idiothetic psychology of personality. In A. Angleitner, A. Furnham, & G. Van Heck (Eds.), *Personality psychology in Europe: Current issues and controversies* (pp. 3–22). Lisse, The Netherlands: Swets & Zeitlinger.

Lamiell, J. T. (1986b). What is nomothetic about "nomothetic" personality research?. *The Journal of Theoretical and Philosophical Psychology, 6*, 97–107.

Lamiell, J. T. (1987). *The psychology of personality: An epistemological inquiry.* New York, NY: Columbia University Press.

Lamiell, J. T. (1990a). Explanation in the psychology of personality. In D. N. Robinson & L. P. Mos (Eds.), *Annals of Theoretical Psychology* (Vol. 6, pp. 153–192). New York, NY: Plenum Press.

Lamiell, J. T. (1990b). Let's be careful out there. In D. N. Robinson & L. P. Mos (Eds.), Annals of Theoretical Psychology (Vol. 6, pp. 219–231). New York, NY: Plenum Press.

Lamiell, J. T. (1997). Individuals and the differences between them. In R. Hogan, J. A. Johnson, & S. Briggs (Eds.), *Handbook of personality psychology* (pp. 117–141). New York, NY: Academic Press.

Lamiell, J. T. (1998). "Nomothetic" and "idiographic": Contrasting Windelband's understanding with contemporary usage. *Theory and Psychology, 8*, 23–38.

Lamiell, J. T. (2003). *Beyond individual and group differences: Human individuality, scientific psychology, and William Stern's critical personalism.* Thousand Oaks, CA: Sage Publications.

Lamiell, J. T. (2010). *William Stern (1871–1938): A brief introduction to his life and works*. Lengerich, Germany: Pabst Science Publishers.

Lamiell, J. T. (2015). Statistical thinking in psychological research: In quest of clarity through historical inquiry and conceptual analysis. In J. Martin, J. Sugarman, & K. L. Slaney (Eds.), *The Wiley handbook of theoretical and philosophical psychology: Methods, approaches, and new directions for social sciences* (pp. 200–215). Hoboken, NJ: John Wiley & Sons.

Lamiell, J. T. (2016). On the concept of "effects" in contemporary psychological experimentation: A case study in the need for conceptual clarity and discursive precision. In R. Harré & F. Moghaddam (Eds.), *Questioning causality: Scientific explorations of cause and consequence across social contests* (pp. 83–102). Santa Barbara, CA: Praeger.

Lamiell, J. T., & Durbeck, P. (1987). Whence cognitive prototypes in impression formation? Some empirical evidence for dialectical reasoning as a generative process. *Journal of Mind and Behavior, 8*, 223–244.

Lamiell, J. T., Foss, M. A., Larsen, R. J., & Hempel, A. (1983). Studies in intuitive personology from an idiothetic point of view: Implications for personality theory. *Journal of Personality, 51*, 213–235.

Münsterberg, H. (1913). *Psychology and industrial efficiency*. Boston and New York: Houghton-Mifflin.

Porter, T. R. (1986). *The rise of statistical thinking: 1820–1900*. Princeton, NJ: Princeton University Press.

Stern, W. (1900). *Über Psychologie der individuellen Differenzen (Ideen zu einer "differentiellen Psychologie")* [On the psychology of individual differences (Toward a "differential psychology")]. Leipzig: Barth.

Stern, W. (1911). *Die Differentielle Psychology in ihren methodischen Grundlagen (Methodological Foundations of Differential Psychology)*. Leipzig: Barth.

Stern, W. (1921). Richtlinien für die Methodik der psychologischen Praxis [Guidelines for the methods of psychological practice]. *Beihefte zur Zeitschrift für angewandte Psychologie, 29*, 1–16.

Stern, W. (1929). Persönlichkeitsforschung und Testmethode [Personality research and the method of testing]. *Jahrbuch der Charakterologie, 6*, 63–72.

Stern, W. (1933). Der personale Faktor in Psychotechnik und praktischer Psychologie [The personal factor in psychotechnics and practical psychology]. *Zeitschrift für angewandte Psychologie, 44*, 52–63.

Thorndike, E. L. (1911). *Individuality*. New York: Houghton-Mifflin.

Venn, J. (1888). *The logic of chance*. London/New York: Macmillan.

Wundt, W. (2013). Psychology's struggle for existence (J. T. Lamiell, Trans.). *History of Psychology, 16*, 197–211. doi:10.1037/a0032319

Jack Martin is Burnaby Mountain Professor of Psychology at Simon Fraser University. His research interests are in the theory and history of psychology and in narrative, biographical psychology, with particular emphasis on the lives and works of George Herbert Mead and Ernest Becker. He is Series Editor for the Palgrave Studies in the Theory and History of Psychology. His most recent book is the *Wiley Handbook of Theoretical and Philosophical Psychology* (2015, with Jeff Sugarman and Kathleen Slaney).

Toward a Social Psychoanalysis: A Conversation with Lynne Layton (Interviewed by Elizabeth A. Corpt, Private Practice)

Lynne Layton, Ph.D., is a nationally and internationally recognized psychologist and psychoanalyst whose work has focused primarily on the influence of culture on the shaping of psychic life, particularly the unconscious influencing of social norms. Her professional career as an academic, psychosocial researcher, prolific writer, editor, faculty member, and psychoanalytic clinician, has spanned over 35 years. Dr. Layton's major contributions have included an innovative bridging of psychoanalytic understanding with academic feminism and the introduction of the often-cited concept of "normative unconscious processes," which was accepted as an entry in the 2014 *Encyclopedia of Critical Psychology*. Her intention to connect psychoanalysis with other academic fields that draw on psychoanalytic ideas has been persistent. It has been influential in the emergent UK/European Psychosocial Network and in her founding of the Psychosocial Work Group

L. Layton (✉)
Massachusetts Institute for Psychoanalysis, Boston, Massachusetts, USA
e-mail: layton@rcn.com

H. Macdonald et al. (eds.), *Dialogues at the Edge of American Psychological Discourse*, Palgrave Studies in the Theory and History of Psychology, DOI 10.1057/978-1-137-59096-1_9

connected to the Psychology and the Other Institute in Boston. Dr. Layton has published over 40 peer-reviewed articles, numerous book chapters, and has co-edited three books. In 2000 she received the Distinguished Publishing Award from the Association of Women in Psychology for her book, *Who's That Girl? Who's That Boy?: Clinical Practice Meets Post-Modern Gender Theory* (1998). Dr. Layton is an active member and current president of Section IX, Division 39, of the APA and is co-editor of the international journal Psychoanalysis, Culture, and Society.

Elizabeth Corpt (EC): Lynne, it's a pleasure for me to have this opportunity to conduct this interview with you. We've known each other a long time now, in various contexts about 20 years. First, I was a candidate in your Gender and Psychoanalysis class at the Massachusetts Institute for Psychoanalysis; then we were co-instructors in the Culture and Psychoanalysis class. We have been analytic colleagues, fellow Board members, and currently we are reading group members. Your work has certainly been instrumental and inspiring to me and to many others. It's no surprise that you are included in this book as a key figure in psychology.

I'd like to begin by putting this interview in some personal historical context. How did you get here? What inspired you to become a psychologist and then to become a psychoanalyst?

Lynne Layton (LL): Well, I actually always wanted to be a psychoanalyst. Becoming a psychologist wasn't really on my mind at the outset. I first had gotten a Ph.D in comparative literature. And then I moved to Boston for an academic job in an English department and, in my second year in Boston, I headed down to the Boston Psychoanalytic to see if I

> could become an analyst. But that, of course, was not possible, because, at that time, academics were not eligible for full clinical psychoanalytic training. This is still largely true for academics without mental health degrees.

After teaching for a couple of years, for various reasons, personal and intellectual, I decided to try to get into the Clinical Psychology Program at Boston University, which, believe me, I was only able to get into because I was on the faculty. And so I got my PhD in psychology. Always, though, with the intention of becoming a psychoanalyst. From the time I was in literature, I was reading and drawing on Frankfurt School critical theory, one of the first attempts to bring together Marxism and psychoanalysis.

EC: So it sounds like your study of the Frankfurt School, in part, inspired your particular path to psychoanalysis, that is, not only focusing on what is psychologically relevant, but also on the social-cultural historical and political context.

LL: Absolutely. My dissertation was on subjectivity in the realist nineenth century novel. So what always interested me was how class, gender and other identity categories are lived in a given historical moment. Largely as a result of reflecting on my own personal struggles, I have always had great respect for unconscious process. And that drew me to the Frankfurt School, to Freud, and to academic writings that applied psychoanalysis to the study of culturally embedded subjectivities.

EC: There's another thread here: your interest in literature, writing, and stories about lives in context. You're a prolific writer. You have four books: the first, *Narcissism and the Text,* co-edited with Barbara Schapiro (1986), and then your book, *Who's That Boy? Who's That Girl? Clinical Practice Meets Postmodern Gender Theory* (1998/2004). Following this, you co-edited *Bringing the Plague* with Susan Fairfield and Carolyn Stack (2002), and then, most recently, *Psychoanalysis, Class and Politics: Encounters in the Clinical Setting,* with Nancy Caro Hollander and Susan Gutwill (2006).

You've written numerous articles. I know, on the PEP Web alone, there are over 40 articles. You've done numerous presentations on gender, post-modernism, social theory, cultural studies. I know that you've also fostered creative collaborations with academics both nationally and internationally. Clearly you've been an influential presence. In general, how have you found the response to your work in the field of psychology?

LL: You know, I live in an uneasy tension with psychology, because I would say that when I trained my department faculty were just about all psychoanalytically oriented. So psychology was, for me, psychoanalytic. And all my clinical placements and supervisions were psychoanalytic.

EC: How rare that is in the current climate.

LL: Exactly. It all changed in the 1990s, just a few years after I received my Ph.D. And those changes, which felt consonant with the intensified individualism that I later recognized to be part of a new neoliberal climate, made it hard for me to continue to identify as a psychologist. I am not sure if and how my work has entered the field of psychology in the United States—maybe it has in the few psychodynamic programs that remain. But my work did find resonance with academic psychologists in the UK, where there is much more interest in what is now called psychosocial studies, and where "critical psychology," an outlier in the field of mainstream psychology, continues to thrive. Some of the founders of the critical psychology movement in the UK, for example, Valerie Walkerdine and Wendy Hollway, embraced my work as did I theirs. That said, my gender book did win the Distinguished Publication Award of the Association for Women in Psychology in 2000, so I suppose my work may have had more of an influence in US feminist psychology than I may be aware of. I know that there are people who do teach my work. But, whether they are in the UK, Europe, or US, they tend to be on the margins of contemporary psychology, people who are left of center, who think psychodynamically and do qualitative work, people who feel that the social context in which we live has to be taken into account in psychological research and in clinical work. In 2010, I had the thrill of working with many of these people on a year-long project at the Center for Advanced Studies in Oslo, titled Personal Development and Psychosocial Change (see Aarseth et al., 2016). Only in a progressive welfare state like Norway might such work be government-funded and even perhaps count as mainstream psychology!

EC: You know, it sounds very much like now with psychoanalysis being on the margins of psychology.

LL: Yes, absolutely. During the short span of my career, psychoanalysis moved from center to margin. And it's interesting, I was recently asked to respond to an article by Otto Kernberg and Robert Michels on what psychoanalytic education should be (2016a). They argue that psychoanalysis ought to ally with university departments. But what they mean by that is that psychoanalysis ought to ally with psychology and psychiatry. When I read it, my first thought was that much of what I have learned about psychology I in fact learned through my studies in literature and other branches of contemporary academic theory, more than what I learned in my psychology courses. Even though my courses were psychodynamically oriented, most of them were designed to create technocrats rather than critical thinkers.

In our current period, psychoanalysis does not have a home in either psychology or psychiatry departments, which are now largely behavioral, symptom-based, and, in psychiatry, pharmacological. But I do think it has a home in the humanities and the social sciences, where meaning is central. So I would want to see the humanities and social sciences become more central to psychoanalytic education.

EC: Yes, I recently read the piece you wrote on Kernberg and Michels. In the beginning of that commentary, you unabashedly claim your outsider status. We've both attended the same institute, and this institute is an outsider, an independent psychoanalytic institute, in that it doesn't belong to the American Psychoanalytic or the International Psychoanalytic Association.

LL: There's a theme here: Outsider!

EC: Yes. In that article you expressed your concern about Kernberg and Michels' suggestion that tightening standards and requirements and making an alliance with psychiatry and psychology would somehow secure the future of psychoanalysis. You made the argument that joining with departments that don't really want psychoanalysis would further denigrate creative and critical thinking in psychoanalytic education.

LL: Yes. I think that's true. I would also argue that literary criticism, or maybe even more particularly sociology and history, have a lot to gain from an alliance with psychoanalysis. In the last 10–15 years, several academics have written memoirs that elaborate on how conflicts around gender, sexuality, race and class are lived. Unlike earlier poststructuralist accounts that celebrated fragmentation and decried any notion of an ongoing sense of coherent selfhood, these accounts reckon with the negative effects of trauma and with the complexities of conscious and unconscious experience, for example, resistance and repetition compulsions. My gender book was aimed precisely at the tension between a poststructuralist gender theory that seemed removed from everyday life and the way I was hearing about lived gender conflict in the clinic.

So, some academics in those disciplines have found that experience from the clinic offers something important to their work and their understandings of themselves. In turn, I have learned from academic theory a great deal about the culture in which we are embedded and about the ways that unequal power relations structure our experiences of gender, race, class, sexuality, a kind of understanding that I feel I rarely get from reading contemporary psychology texts.

That said, I have to admit that I seem to have been unconsciously motivated, given my family history, to try to bring groups together that are not speaking to each other but that clearly, in my mind anyway, ought to be speaking to each other. So, you know, at the core of *Who's That Girl?* was this train of thought: there are these fascinating post-structuralist thinkers, like Judith Butler, but sometimes what they say about people makes me wonder, have they ever actually talked to a person? But then, on the other hand, as you know, because we've worked together to counter this trend, what's idealized and fetishized in the clinical world is a psyche outside of any social embeddedness, a universalistic psyche. And from that kind of perspective, which is often rationalized by the need to maintain confidentiality, it "makes sense" to do such things as change a person's gender or age in your case presentation. And that suggests that the treatment offered was oblivious to the particularities of the patient's gender, age, general social positioning.

EC: Yes, right.

LL: So yes, so I do feel like the two camps need to talk.

EC: This has both a personal psychological meaning for you, as well as a broader intellectual meaning, to have groups that don't usually communicate or cross-fertilize, do so, or at least attempt to.

LL: Over time, the personal psychological meaning has become more apparent.

EC: I see, I see.

LL: The disciplines are still not talking to each other, so I keep trying.

EC: Shifting gears a bit, I want to ask you about your work on normative unconscious processes, because that is one of your major contributions to the field. I'm going to quote you here on your definition. "It's the effect of the workings of unequal power arrangements on identity formation and relational interactions." Your clinical examples of this, in particular, are quite compelling in explicating the psychic cost, emotional pain and ruptures of intimacy that result in living these splits.

To my mind, this contribution has been invaluable in coming to understand the way we split our identities to conform to what we have deeply unconsciously internalized as proper or improper in relation to the identities that we live and their place in the social hierarchy. This concept has certainly been key to me in understanding and formulating my own experience of class, and in thinking about class and forming a psychoanalytic identity.

Overall, I'm curious, what have you felt to be the field's response to this concept? And we can think about that in terms of the psychoanalytic and psychological field's response to this concept. How do you understand the reaction? I recall when you and I taught Culture and Psychoanalysis and you were introducing these ideas, as we were trying to introduce ideas about culture in general. We always encountered a certain resistance amongst candidates who kept asking, "Why do we need to think about this? Why do we need to talk about this? Can't we have another course on Winnicott instead? Isn't this a waste of our time?"

We were curious about what that was about. Is the concept of culture so large, or so deep, that it's very hard to access and to make sense of its impact? Luckily, in our institute, we've now moved the Culture and Psychoanalysis class from the third year (or third class position) and placed it in the first year of training. The intent is to encourage candidates to begin thinking about those ideas from the very beginning.

But back to normative unconscious processes. What has been your experience over time in the response to this concept? Have you encountered resistance to this concept, and if so, what sense do you make of it? Or has the concept been accepted, and incorporated?

LL: Let me start by saying a little bit about the concept of normative unconscious processes and how I came to it. In *Who's That Girl?* I was arguing for the importance for clinical work of post-structuralist critiques of identity categories and the power hierarchies that those categories support. However, I was also arguing, against post-structuralist accounts, that categories such as gender, race, class can also have positive psychological functions, can facilitate creativity and counter-hegemonic resistance. Afterwards, I began to recognize that what psychoanalytic work tends to ignore is the way these categories and culturally supported hierarchies get in the way of health and creativity, how cultural inequalities are repeatedly reproduced psychically. So, yes, I think my contribution to clinical psychoanalysis has been to focus on the many ways in which therapists unconsciously reproduce an unequal status quo, or where patients and therapists together collude to do so.

Now, as for the concept's impact in the field and resistance: even as we were teaching this course together, maybe shortly before or at the same time, I was teaching culture and psychoanlaysis in multiple places. I was teaching it to undergraduates at Harvard who were social studies majors and women's studies majors. I was teaching it in the internship program at Beth Israel, to psychology interns.

EC: Yes.

LL: I was teaching it at the institute. Oddly, I was thinking about this question of impact this very day, because I just came back from the 2016 Division 39 spring meeting, where I again discovered that graduate students tend to be very interested in my work. But students who are more advanced in the field seem to have much more resistance to the ideas. Parenthetically, I should say that I'm not a very professionally political person, not the kind of entrepreneurial self one is accustomed to in this day and age. So I haven't done much to "brand" my ideas. And I have not found much of a professional psychoanalytic community here in Boston that might share my perspective. So I write, I teach, I give talks when invited: when people are interested, they're interested. But, as I said earlier, the kind of work I do runs quite counter to trends in the fields of both psychology and psychoanalysis today.

So I definitely have noticed, over time, that the less engagement with psychoanalytic education someone has had, the more meaningful what I talk about seems to be to them. Clearly, one possible reason for this could be that, as you mature in the field, you find my ideas are not deep enough, that they're not helping you deepen your clinical work. But it is also possible, and this is something that came up in the recent conference, that as one gets indoctrinated into the field's dominant discourse, as you learn what it means to be doing "proper" psychoanalysis, as you take part in an education that offers only one course in culture, if that, you soon come naturally to feel that what you really need in order to go deeper is another course on Winnicott. It's always this concept of—

EC: —deeper, yes.

LL: —deeper.

EC: So, as people's identities start closing in and becoming more rigidified, in some way their ability to stay open to forces larger than themselves may be impacted.

LL: When it comes to professional identity, absolutely. Some of the difference in openness to these ideas has to do with discipline. If you've been educated as a clinical social worker, for example, I think you might be more aware of the social embeddedness of the psyche—because of the discipline's ethos as well as the kind of populations you work with. I think psychology training—and this is partly why I don't identify as a psychologist—is very individualistic. I used the word technocratic before, but it's also a field that is very, very focused on the individual outside of any social context.

EC: So I wonder whether those who have more skin in the game, you know, the ones that are most affected by "isms", the people most directly struggling with the impact of class or race or gender, have to pay attention to these things. You then find this material relevant.

LL: Yes, that is also true. I know we talked about this when we were teaching together: sometimes there might be a female candidate who had earlier taken the course on gender and had not seemed to find it relevant in any way to her own psychic issues or her clinical work. Later, perhaps, she experiences an ugly divorce or becomes more aware personally of sexism. And suddenly, the ideas become alive for her. Gay and lesbian identified candidates, have also, in general, found the courses I teach to be "deep." It is often those who are "unmarked," the white, privileged heterosexuals, who find talking about gender, race, class, and culturally-imposed traumas irrelevant. Many young clinicians have told me that their supervisors and teachers train them to stop thinking of this kind of material as important and to stop bringing such material into their case presentations.

And I remember people in our class dismissing the articles we were reading with comments like, "Oh, this person is talking about class just to find a niche," as if they simply could find no relevance in this kind of thinking. I am always struck by how many layers of psychic work must go into the disavowal of the claim that class means something, that, as Joanna Ryan (2014) says, class is "in" you.

EC: That's right. I find that the clinical examples you use in your writing are quite powerful, in terms of how you can listen and hear your patients touch upon your privilege, and the way you can bring it back into the session. It's so easy to overlook those things and simply just be comfortable in one's privilege instead of problematizing it and allowing it to be a part of the analytic process.

LL: Yes. I often give talks in which I focus on becoming aware of my own unconscious process as it becomes enacted in a normative repetition of sexism or classism or racism. And while some have greatly appreciated how vulnerable I am in my written and oral work, others can be quite dismissive. I will hear things like: "You're bringing politics into a place where it ought not be, and it isn't relevant to psychoanalytic work." Or, and this one really irks me because it is the very antithesis of my intent: "Wow, you make a lot of mistakes, don't you?"

EC: Yes, I remember reading that in one of your papers. Someone in the UK made that comment. Tell me what that was like for you?

LL: It was quite shocking, because when I first started writing about normative unconscious processes, I used other people's published clinical work, which, as you can imagine, did not go over very well.

EC: I remember that, yes.

LL: So then I started using my own. My hope was to illustrate how we are all largely unconscious of the ways in which we are positioned within cultural hierarchies of gender, race, class, and sexuality. We are unconscious of how we subtly prescribe our view of what a "proper" identity is. The "You make a lot of mistakes" response basically translates to "You're not a very good clinician."

I imagine that the person in the UK who said that was trained in a Kleinian or classical tradition where the analyst is often figured as someone who is supposed to be a perfect container and not act on those kind of transference/countertransference feelings. I obviously don't see them as mistakes but rather as somewhat inevitable consequences of being positioned in various cultural hierarchies and ideological positionings.

EC: Yes.

LL: I come to this position from a relational psychoanalytic tradition that acknowledges and problematizes the fact that the analyst's unconscious is always a part of the work.

EC: Yes, sure.

LL: If the unconscious is shaped by social factors, and there are two unconsciouses in the room, then these kinds of enactments that unconsciously contribute to reproducing a sexist or racist or classist status quo are probably quite common. As I've argued, they tend to reproduce what has made the patient ill in the first place. I talk about them to help clinicians become aware of these kinds of transference/countertransference moments so that, perhaps, we just might be lucky enough to catch them when they are happening (Layton, 2000a, b, 2002a, b, 2004b, 2006a, b, c, d, 2007, 2008a, b, 2010b, 2011b, c, 2013a, b, 2014b, 2015, 2016b, 2017; Layton, L.B. 2014).

EC: Yes, it seems to me that staying open to the influence of one's own normative unconscious processes, and one's own inevitable blind-spots, does leave the analyst or therapist vulnerable in a certain way; in a good way. From your perspective, striving to be a perfect container reads as an attempt to insulate oneself in a kind of privilege. This is the opposite of your project.

LL: Yes, I most definitely hope that my work on normative unconscious processes will add to the rich relational literature on the analyst's vulnerability, particularly highlighting the invitation afforded by the clinical setting to hide those vulnerabilities behind our privileged position in the treatment.

EC: If I might switch gears again, I'd like to bring us around to thinking about your recent work, which is focused on the psychological impact of neoliberalism, particularly the way neoliberalism induces the kind of trauma and disruption of trust in the social world, a social world that should take into account the needs of all people, both the vulnerable as well as the powerful. I wondered if you could say something about how you began thinking about neoliberalism as a serious area of study worth your attention as a psychoanalyst.

LL: As I was saying earlier, I think I have always been bothered by the individualism inherent to the discipline of psychology. And I suppose I probably first just thought about it as something inherent to the field rather than as a phenomenon that changes historically. And the first time that I began to think about individualism as a phenomenon with a particular historical inflection was possibly in my 2009 article (Layton, 2009), "Who's Responsible? Our Mutual Implication in Each Other's Suffering." And I'm not even sure that I had the word "neoliberalism" then, because it's not a term that is used in public discourse. Neoliberalism, I've come to understand, is many things: it's an ideology; it's a political and economic system; indeed, it permeates every institution such that its effects can feel "normal," like "common sense"; it's the oxygen that we breathe right now.

EC: Right, sure.

LL: Actually, and still at a time before I had the word neoliberalism, I believe I first started thinking about an intensified form of individualism in relation to post-9/11 politics. There was a conference called "The Desire of the Analyst," and my paper was on what I called the politics of attack and withdrawal (Layton, 2006c). I was starting to see different group formations emerging in the United States. Some groups were going after the most vulnerable among us in a politics of attack; another formation was marked by withdrawal from politics and a search for respite in the private sphere.

Then I just started reading what academics were saying about our contemporary world. I learned what I know about neoliberalism not only from the clinic but from exploring other disciplines. Again, I tried to bring the two sets of discourses into conversation (Layton, 2004c, d, e, 2009, 2010a, 2011a, c, 2013a, b, 2014a, c, 2016c, d).

EC: Sure.

LL: In academia, the concept of neoliberalism has been written about since the 1990s and throughout the 2000s. UK thinkers had actually been writing about it even longer because they recognized that Margaret Thatcher's program of de-industrialization and her skepticism about there being any such thing as "society" had inaugurated something new in the realm of both economics and social policy. Miners were striking, and the working class was suffering from her policies. Many vocal UK critical psychologists and sociologists are of working class origin and so were acutely aware of the class nature of these changes.

EC: That's interesting. It seems like social class is so much more addressed in the UK.

LL: Yes, for those academics originally from the working class, class never fell off the research agenda, whereas in the US, class was largely replaced by a focus on gender, sexuality, and race difference. One could say that it is an effect of neoliberalism that class became nearly absent from the US academic research agenda.

EC: Yes. Those working class academics had skin in the game and brought their experience into the academy as an area of study, to the dismay of those who wanted it off the agenda.

LL: And although class is an important part of my own identity formation, my earlier work, too, focused first on gender, then on sexuality, then race. Somehow class disappeared, mowed down in the neoliberal claim, after the fall of Communism, that there was no alternative to capitalism and that giving free reign to market dynamics would lift all boats. Meanwhile, slowly, inequality was proceeding apace until it smacked the public in the face during the Occupy Movement. I continue all the time to expand my awareness of neoliberalism's pervasiveness. For example, I don't know if I had told you about this before, but I remember giving a talk at APCS, the Association for the Psychoanalysis of Culture and Society, on neoliberalism. And an African-American friend of mine, not realizing he was commenting on my own paper, later said, "I don't know how anyone can talk about neoliberalism without talking about race." At that time, I didn't really know what he was talking about. And then I read Michelle Alexander's (2010) *The New Jim Crow*, which is about mass incarceration and the war on drugs, both of which have devastated large parts of the African-American community, creating a relay between what Loïc Wacquant (2001) has referred to as the neoliberal "hyperghetto" and the neoliberal prison. And that made me able to hear in a new way something a patient of mine had said some time before: she said she felt her parents had consciously and unconsciously delivered a strong message to her when she was growing up, and that message was Yale or Jail (Layton, 2016c). And that "Aha" moment became the basis of a new paper I wrote on neoliberalism, which incorporated the different ways neoliberalism is lived depending on the intersectionality of one's class, race, gender, and sexual positioning.

EC: You can't continue to think the same way once that light bulb goes off.

LL: Exactly. And then you start hearing it in your work with patients, you start reading things about it. For example, there is a wonderful paper I have been recently quoting a lot, by Rosemary Rizq (2015), in which she talks about how neoliberal policies and what she calls "the audit and surveillance culture" have infiltrated the UK National Health Service, such that clinicians feel as though the increasing amounts of paperwork they have to do become experienced as equivalent to giving care. She illustrates this via a complex and unconscious institutional enactment that began when a patient complained about having been treated as a number and not a person.

EC: In this country, there is enormous pressure on clinicians who serve on insurance panels to focus on data, evidence, efficiency. The clinical and certainly the context gets overlooked and minimized, but at great cost. Overlooking the socio-cultural context and focusing on the individual as an isolate is always problematic. I know a number of years back, you wrote a paper in which you questioned whether we, as psychoanalysts, were simply making better narcissists out of our patients (Layton, 1998, 2005a, b, 2006a).

LL: Yes. My contention in several things that I have written is that if you leave the socio-cultural realm out of your understanding of your patient and yourself, you are at risk of creating healthier narcissists. In the chapter of *Who's That Girl?* on Kohut, I critiqued a tenet of the self psychology of that era, that the developmental line of narcissism was separate from the developmental line of object relations. I was looking at the papers published in one of the early self-psychology case books, and I felt that the patients, at the end of the treatment, were healthier narcissists who were not treating their significant others much better at the end of treatment than at the beginning. The theory seemed to have led to a lack of concern about how the patients were treating others. Around the same time, I read some papers by Phil Cushman (See this volume), a hero of mine in the field of psychology, someone who always looks at the social roots of our psychic theories and psychic woes. Phil was looking at self-psychology's concept of an empty self as a historical phenomenon, and he argued that the theory and proposed treatment took as normative a consumerist self. And yes, I think that was partly what led me to the idea about the risk of treatment creating healthier narcissists. To me, an ethical psychoanalysis has to aim at expanding the capacity for what Jessica Benjamin (1990) has called "subject-subject" relating, mutual recognition (Layton, 2002b, 2004a, b, 2005b, 2013b).

EC: Well I wonder, given what you were saying earlier about the way psychology departments have been co-opted by the need to bring in funding, evidence-based research . . . —

LL: Can I just add something here? I just thought of something that pertains to the last question.

EC: Yes, please.

LL: So part of what—part of what the "Who's Responsible?" paper was arguing, and this is, I think, really a challenge for clinical work, is that if you think about the ways that identities are built in relation to each other, and that they're built in a particular socio-historical moment that is shadowed by long and unconsciously transmitted earlier histories, I think you begin to see the complicity in each other's suffering that I was trying to get at. And that speaks to what you were saying earlier about the therapist's privilege. The healthy narcissist is addressed as one who is not complicit, who simply needs to build a better self, a more actualized self, one that has little bearing on other relationships and other social groups.

Now it would be tricky to bring this awareness into the clinical arena. But I think there are moments when it is possible. For example, in the "Who's Responsible?" paper I talk about a boss who realizes she needs to make time for herself, to stop working 24/7. But she was still expecting her workers to work those kinds of hours. So where is her empathy in that moment? How might a therapist foster empathy in such a situation? This patient was generally a very empathic and lovely person but empathy sometimes failed her when it came to the pursuit of making money and running an "efficient" business. I was arguing in that paper that we seriously have to rethink our conception of empathy if increasing empathic capacity is central to the ethic of psychoanalysis. Empathy is a degraded concept if limited to use in only certain situations. My argument was that we should aim for an empathy that evolves from recognition of complicity in power relations, one in which the privileged are permanently altered in their encounter with the less privileged.

EC: It's actually a bit ironic. Self-psychology, or at least classical Kohutian self-psychology (1971), emphasizes the importance of self-objects to realize a healthy self. But yet, in some way, thinking about the project of maximizing one's fully realized healthy narcissism implies that the need for self-objects, at some point, falls away.

LL: Yes, yes.

EC: But you need a village to be a self. You need something communal to be a self.

LL: I completely agree.

EC: But yet, there is this tension about being one's own person.

LL: Yes, and I want to ask you a question that relates to what you're saying. I know self-psychology has developed quite a bit since the classical moment. But in the classical moment, which is when I wrote my psychology dissertation on Kohut, it felt to me like the self-object was understood as devoid of any function beyond the psychological nurture of the self. I know people have talked about self-psychology as a one and a half person theory.

EC: Yes. Yes.

LL: And that was what it seemed to be in my reading of the literature. It did not seem to be aiming at developing the kind of capacities that Jessica Benjamin (1988) speaks about as mutual recognition. But self-psychology may well have developed beyond that early phase.

EC: Self-psychology has developed beyond looking at others as self-objects providing self-object functions. There has been the turn toward intersubjectivity, where, you know, Stolorow and Atwood (1994) talk about mutual influence always being at play, even preceding the capacity for mutual recognition. They refer to the "unbearable embeddedness of being" (p. 243), meaning that somehow we are always and forever implicated in each other's lives. We are far from being psychological isolates. But I think now, in the swing toward the ethical turn in psychoanalysis, with the introduction of philosophy, particularly that of Levinas, we, in some way, have flipped things around in the extreme. Now, according to Donna Orange, the individual is actually held hostage by the other. So in some ways, it's a huge correction from self-objects providing functions.

LL: That's really interesting. And that's called a self-object also?

EC: No. The language of self-object has completely fallen away. It's not even used anymore.

LL: Thank you for illuminating that shift for me, the swing to another extreme. Because that touches on another area of contemporary theory with which I've often felt in tension. It comes up when I read Judith Butler, for example, the continental idea of "the other" as oppressive, which appears from Sartre on, if not before. That feels quite problematic to me, too. When you look at Butler's *Giving An Account of Oneself* (2005), for example, which we read together in our wonderful reading group, you find that others to the self are frequently figured in war-like terms, as intrusive, impinging, attacking.

EC: I'd love to do some more thinking with you about this subject at some point.

LL: I'd love that, too.

EC: Good. So, getting back to neoliberalism, I'm wondering, then, whether we as psychoanalysts are blinded by our own traumatization and disruption in the face of neoliberalism. You use a term in one of your papers, where you talk about a "group-enforced denial of dependence and interdependence." I wonder, in the fields of psychology and psychoanalysis, about this race to survive in a hostile climate.

LL: Yes, it's so much bigger than the individual.

EC: Yes, so much bigger.

LL: The individual, you know, you can see it in the kids, in the whole "Where am I going to get into college?" thing.

EC: Yes.

LL: Or, beginning with the parents' anxieties: "Where am I going to get my kid into preschool so s/he can get into the right college?"

EC: Yes.

LL: You know, I've had patients say, "I know it is crazy to get caught up in this, but you get caught up in it or die."

EC: Yes, as you said earlier, "Yale or jail."

LL: Yes, my patient's parents didn't have to say it, but you know, the feeling came through: succeed and be part of the one percent, or be a loser. It's just a terrible atmosphere to grow up in. I see so much damage too, so many symptoms that are bred of that pressure. But it would take a movement, a social movement, to contest it.

EC: Yes, yes, that's right. So I wonder, how do we find our way out of this blind spot as a field?

LL: That's a very good question, and it came up many times at the recent 2016 Division 39 Spring Meeting. The 2017 meeting has the theme "The Times, They Are A'Changin'. How About Us?" At many panels that I attended or spoke on, people in the audience, mostly, but not all young people, asked the question, "How are we going to get psychoanalysis to change?" The diversity of gender identification, of sexuality identification, of ethnicities and races at that conference was exciting. And the conversations were exciting. Maybe they will change the field, I don't know. But at the moment, for me, the meaning of impossible profession includes a sense that clinicians are called upon to treat socially-created psychological problems that go unnamed and unclaimed as social problems. For instance, I recently read an article that talked about the rising suicide and suicidal ideation rates in young people across Massachusetts. The author briefly mentioned an increase in pressure to succeed, but the article's main focus was on new collaborations between community mental health centers and schools that are helping ill and previously hospitalized students re-integrate into the schools. My first thought upon reading it was, "And who is going to attend to the stress of the clinicians who are being called upon to "fix" what is actually a social problem spawned by neoliberalism?"

That said, I think that there are openings for change in the field, openings that could indeed lead to social activism as well as better clinical work. I feel hopeful because there is a group of us, including you, who continue to write the kinds of papers that might bring change. I'm certainly not alone. Indeed, I was only able to elaborate the concept of normative unconscious processes after having read many papers that described the kind of enactments to which the concept refers—works by Altman and for example. These papers, your paper, "Peasant in the Analyst's Chair," on class (Corpt, 2013), are very meaningful to younger clinicians. And hopefully to some seasoned clinicians as well.

EC: So when I think about the caretaking role that we are in, in the consulting room, and you've addressed this in the many clinical examples you've given, I also see that we have a caretaking role in a larger way, in a larger systemic way. But I wrestle with how to speak to that beyond our being in our individual consulting rooms?

LL: As it happens, I feel like I'm in a new phase of my life that I probably was not able to initiate while I was doing clinical work and teaching. Now I find myself doing more mentoring type work, some of which involves work outside the clinic. I've just founded a new group in Boston that is based on a model that was developed in San Francisco, called Reflective Spaces/Material

Places. The idea is to gather together people who are doing community mental health work or who have an affinity toward community mental health work, and who are also psychodynamic in orientation. These people are, in a way, assailed on two fronts. There is no money for community mental health, and very few clinics now encourage or teach psychodynamic work. And sadly, the Boston Institute for Psychotherapy has just—

EC: —has just announced that it's closing. Yes.

LL: That's really a tragedy.

EC: It's a death, it's a death, really.

LL: It's absolutely a tragedy, one of the only psychodynamic training sites left and one of the only places where people who can't afford private practice fees can get good psychodynamic long-term treatment. So we're hoping that Reflective Spaces/Material Places—Boston will be a space in which we can all reflect together on the relation between clinical work and what's happening in the larger culture. We'll talk about the kinds of things that come up in clinicians' work with the very diverse populations that they see, what kind of care the most vulnerable need to get and what they are getting. So I feel like I'm entering a more mentoring and activist phase. And I've also just become President of the Psychoanalysis for Social Responsibility Section of Division 39, Section IX. And I'm very excited, because there are lots of young people there interested in social justice and activism. That's partly why I became President, because I wanted to work not only with my devoted colleagues of many years but also with the wonderful young people on our board and membership; they are so thoughtful about the issues that I have dedicated myself to in my writing and teaching career. And I think they're carrying on the work. And it's just wonderful.

EC: It's really the future.

LL: Yes, I hope so.

EC: Reflective Spaces/Material Places is such a wonderful idea. I wonder whether you might imagine that, at some point, such a forum could be open to the general public, as well as to the professionals in the field. I'm not even sure how to do that.

But it strikes me that we somehow need to reach beyond our own borders to somehow bring that space for reflectiveness to a wider audience. When I think about what's happening in this election process, and think about the emotional suffering that's being played out before our eyes, is there some way for us as clinicians, as people who think about these ideas deeply, to reach beyond our offices, where only a limited number of people can be seen?

LL: That's such an important idea, Elizabeth. I have noticed recently that there are some psychodynamic psychotherapists and psychoanalysts who are writing for the *New York Times* and other media outlets, and they are indeed making a case for thinking psychoanalytically about what is going on politically. I was part of a group last year that did a couple of presentations that we called Manifesto Fest.

EC: Oh yes, yes. That was great.

LL: Our aim was to try to speak in accessible language about the value of psychoanalysis, and some of us specifically spoke to its value in understanding social problems and what goes on in the public sphere. My own contribution focused on my perception that several mainstream journalists and op-ed writers who were trying to understand, for example, police violence against black men, had begun to point to the crucial role of unconscious process, particularly unconscious racism and unconscious sexism.

EC: Yes. It's been showing up more and more in the press—

LL: So this might be the moment for psychoanalysis to make a cultural contribution outside the clinic. I think these articles by clinicians are getting accepted by the press because people realize that what is going on defies explanations that focus solely on facts and logic.

EC: We're reaching the conclusion of our interview. So here is a big question. What do you see as the pressing historical/political/theoretical concerns for psychology or psychoanalysis today? What will have the most influence on the future of what we do?

LL: I wish psychology—I would wish for psychology to engage more with sociology, social science—and this would be a very big change, because sociology departments are generally as hostile to psychology as psychology departments are to sociology. We have already talked about what psychology could gain from thinking sociohistorically. The reverse is also true. Sociological studies would greatly benefit from an encounter with psychoanalysis. A study that comes to mind is Jennifer Silva's (2013) book *Coming Up Short*, which elaborates the effects of neoliberalism on the young black and white working class adults she interviewed. Her results were fascinating and very psychological, but I felt that the study could have had even more depth had she gone beyond self-report and taken unconscious process into account.

I would love to see the university become—and I have said this, now, for 30 or 40 years—more interdisciplinary. I think that is really where the hope lies for psychology. The fragmentation of disciplines very much leads people to put too much effort into sustaining their own field's status, for example, defending one's own department against budget cuts.

EC: Yes, well it's part of neoliberalism to pit departments against each other, and different branches of clinical work against each other.

LL: Yes, exactly. And my whole career has taken place in the time of neoliberal thinking and cutbacks, although I came of age during the very short period in US history of relative equality and broadly shared concern for the most vulnerable. It was in the moment where there was felt to be plenty that interdisciplinarity seemed to have flourished and, luckily for me, that ethos permeated the period during which I was getting an education.

EC: Interesting.

LL: There are some spaces, like the American Studies Association, for example, where people come together from different disciplines. The ASA conference I went to in 2015 was one of the most exciting and innovative conferences I have ever attended.

EC: Yes, that's the best.

LL: That's the best, and it holds the most hope for the future of psychology and for everything else. You could see there the impact psychology and psychoanalysis could have if they aim beyond the isolated individual. In that spirit of interdisciplinarity and social-mindedness, I am very proud of the journal that I have now co-edited for twelve years, *Psychoanalysis, Culture & Society*. It is an international, interdisciplinary journal that embraces all schools of psychoanalysis, academic and clinical, publishing work that fits our mission of addressing the unconscious roots or consequences of social problems, inequalities, and injustice.

EC: Historically, the practice of psychology has not associated itself with any specific forms of political activism or social responsibility. Although even as I say that, I recall that you began this journey with the Frankfurt School. So in your work, you have gone about things quite differently. How do you understand that? And what inspires you to hold to a broader and more encompassing way of thinking and practicing?

LL: Well yes, my initial contact with psychoanalysis was in academia, where I encountered it as a revolutionary theory and practice, one that challenged externally imposed as well as internalized sexism, for example. Psychoanalysis was so important to my experience of what it meant to be feminist, so important to understanding resistance to change, the repetition compulsion. I read Nancy Chodorow in comp lit graduate school, a wonderful example of the rich possibilities of bringing sociology and psychoanalysis together. The Frankfurt School, especially Fromm and Marcuse, in very different ways, helped me understand the relation between subjectivity, character, and socio-historical change. Adorno and Horkheimer's (1944/1972) *Dialectic of Enlightenment* was crucial for me in understanding the historical psychosocial roots of modernity. Freud's psychosocial writings are still groundbreaking—another of my suggestions for improving psychoanalytic education is my feeling that we would greatly benefit, as institutions and as individuals, from learning about group unconscious process. We learn very little to nothing about this in our education, another way the field sidelines the psychosocial. Learning about group process would help us understand how we're interrelated. It would help us understand our longing for leaders, how we yield autonomy, how what we think we're doing consciously as a group can get subverted by unconscious desires that run counter to accomplishing the task. Bion's work on groups is so important here, and it is rarely if ever taught. Our education does not privilege the revolutionary edge of psychoanalysis.

When I think of that revolutionary spirit, I think of George Makari's (2008) book, *Revolution in Mind,* Wilhelm Reich and the Frankfurt School, Russell Jacoby's *Social Amnesia* (1975) and *The Repression of Psychoanalysis* (1983), which, again, I read as an academic before I was a clinician. I think of Jessica Benjamin's (1988) *The Bonds of Love* and Juliet Mitchell's (1974) *Psychoanalysis and Feminism.* It feels like my early training as an academic was invaluable to me in giving me a particular sense of what psychoanalysis and psychology could offer when connected to a desire for social justice and radical social change.

EC: You know, I think you almost have to step outside of psychoanalysis to see the revolutionary aspect of it. And there's so much emphasis on belonging in a certain way within the halls of psychoanalysis, that we forget the revolutionary roots of it, and the political roots of our early forebears.

LL: Yes. You were asking me before how I think neoliberalism has affected our field.

EC: Yes.

LL: We too have not been able to avoid the push to become *Homo Entrepreneurs*. We, too, have to brand ourselves to make our way. We are not immune to the pressure to make our name on some concept—like normative unconscious process. We wonder how many people cite it? Cite me? And all of that works against collective action and change.

EC: What do you see as the crucial change to your way of thinking and your approach to the discipline over time? I mean it sounds to me like it's been a kind of a building up and a sedimentation of ideas—

LL: Yes, I think that's accurate. For example, here is how the idea of normative unconscious process began to develop in my mind. At first, in the early 2000s, because my work was focused on gender,

> I called it a heterosexist unconscious (Layton, 2002a). Meanwhile, I continued to read what academic feminists were writing. And the focus started to shift in the'90s toward understanding identity in terms of intersectionality, for example, the recognition that race is always classed and gendered, gender is lived differently depending on one's class and race.

Then, responding to my work on heterosexist unconscious, analyst Stephen Hartman wrote a paper on what he called class unconscious, how he and his patients live class and the way class marks their clinical encounters. Then I wrote something on race unconscious, which brought together academic writing on race with work on what they call "racial enactments" by Neil Altman (2000), Kimberlyn Leary (1997), and others. But it was at that point, I think, that I started to recognize that thinking intersectionally is the best way to think about identity and identificatory processes. I began to realize, too, how different subcultures create their identities in relation to dominant norms of gender, race, class, sexuality. And so that's where I shifted and began to think in terms of normative unconscious processes.

Then I began to see it all in the context of neoliberalism, and I began to wonder what kind of normative unconscious processes are we enacting that contribute to reproducing a neoliberal status quo? Class, of course, is crucial to thinking about that. But again, the way that class is lived in neoliberalism is different depending on race, gender, sexuality. And now, one thing I've been thinking about recently, and that I have written a little bit about, is technique. There is a group of clinicians who are thinking and writing about the importance of history in understanding subjective experience, the importance of collective identifications, the importance of ideology, and the importance of how we, as clinicians, can be complicit in sustaining an unjust status quo or crucial in changing it. I'm curious about how thinking about all of that might change our thoughts about technique. I read something by Eyal Rozmarin last year that made me start to think about that, and then I began to look for hints at questioning technique in the papers of those who write about socio-cultural matters. And I found some: Katie Gentile, for example, has interesting thoughts about broadening our concepts of empathy, about which I have also written. Orna Guralnik's writings about ideology and interpellation also point toward changes in technique. Sue Grand's work is as deep and as psychosocial as it gets.

So I am just at the point where I've gathered some examples of how thinking psychosocially might affect clinical technique. Another is Françoise Davoine, who was a keynote speaker at this recent Division 39 conference, and who has also contributed, with her late husband, Jean-Max Gaudillière (Davoine and Gaudillière, 2004), to a rethinking of technique. Like relational theorists, they understand their work with their patients to be co-constructed; they call themselves and their patients "co-researchers". They consider their historical biographical subjective experience to be a part of the treatment that, when it becomes conscious at crucial moments, can be transformative in helping the patient find his/her place in history, helping to restore what they call the broken link to the social that has caused them to fall ill. In the course of their work, for example, they might tell their patient one of their dreams, because they consider dreams about the patient to belong to the treatment.

EC: Yes, they're much more egalitarian in terms of how they are with their patients.

LL: Yes, because they feel that their own histories are inextricably intertwined with the histories of their patients.

EC: Well my final question was along those lines, about where you imagine your future work will take you. And it sounds like you're already headed into some fascinating areas.

LL: Yes, the question of technique is one.

EC: Are there others?

LL: Well, I'm curious to see where these more activist things will take me. One place they are already taking me is to greater awareness of group conscious and unconscious process. I have started several groups in the past two or three years, and have taken leadership roles in, for example, Section IX. And what I find is that in the course of simply trying to run a meeting so many other things are happening that, at first, feel like digressions from the task. But sometimes the way those group processes emerge and are dealt with brings you precisely in touch with the revolutionary nature of psychoanalysis. On more than one occasion I have felt that pausing to process what is happening in the group brings me closer to the goal of being able to walk the walk rather than just talk the talk. It can be frustrating, but it's also exciting. I feel I'm going to learn a lot from group process—and from being in groups with people, young and old, who are committed to connecting psychoanalysis with social justice.

Last year, I participated in a group relations conference, which was a fascinating introduction to the manifold ways in which unconscious process is social. While I had some issues with the group relations concept of authority, it did help me better understand how groups work, consciously and unconsciously. Which reminds me—and this brings us full circle, back to your question about my relation to the field of psychology, I actually am now teaching, for the first time, in a psychology department, a rather revolutionary one.

EC: I see, where?

LL: At Pacifica Graduate Institute. I'm teaching in the PhD program in community psychology, liberation psychology, and ecopsychology, and my course is called social psychoanalysis. I devoted one of the three day-long classes to Freud's basic concepts, in socio-historical context: transference, repetition compulsion, resistance. And I taught some of his social papers and books. The second day I taught several different paradigms that focus on group unconscious process, including a fascinating foray into a field called Social Dreaming. The idea is that people in the group's individual dreams, with the group's associations, reveal something about the social world and context in which they find themselves. On the third day we looked at themes such as neoliberalism, ideology and identity formation, and intergenerational transmission of trauma. There I taught a variety of papers, including your paper on class. The students mostly already work or will work with traumatized and oppressed populations, and the program is committed to teaching the significance for such work of understanding the facilitative and regressive effects of unconscious process. Although the students did not come in with a positive feeling about Freud, many of them grew greatly to appreciate his work and its impact on all the work I taught. Teaching this course, in this department, with these students, was something of a career highlight for me.

EC: That sounds so exciting.

LL: Yes, this department, as well as their clinical psychology department, is committed to teaching the connections between social justice and psychosocial work. Sadly, to return to our earlier theme of neoliberalism, the Pacifica Clinical Psychology department, which is all psychodynamic, was recently turned down for accreditation by the APA. My understanding is that a reason given was that it was not "evidence-based" enough.

EC: Well, as I listen to you tell this wonderful story, right where my mind went is, "Will these people get jobs once they graduate? Or will they tilt toward needing to become psychoanalysts, with the hope of making a living?"

LL: I do think the clinical students might have trouble getting into APA accredited internships in some states, which could affect their job chances. The students in the program I teach in are not being trained primarily to be clinicians, although some of them already are therapists. They're being trained to work with groups, for example, the homeless, immigrants, people seeking asylum, and in educational settings.

EC: We need this, yes, yes.

LL: But, as I said, there is a Clinical Psychology and Counseling Psychology program there, and I will do some teaching for them next year as well. The chair of that program is a tireless advocate for depth psychology and social justice and has been critical of the American Psychological Association for their starring role in marginalizing psychodynamic work and in construing evidence-based practice very, very narrowly. And, in fact, the evidence has begun to mount in favor of the superior efficacy over time of long-term psychodynamic work.

EC: Yes, the research has shown that.

LL: So what are bodies such as APA going to do when they are faced with this evidence? They've been demanding evidence-based treatment all these neoliberal years, and then the evidence starts to show something that insurance companies aren't going to like—because it's expensive. That will be an interesting challenge to neoliberalism.

EC: Yes, that will be very interesting. I also love, in one of your papers, you wrote about Kim Leary's wonderful idea that, instead of a third analytic training case, people could be encouraged to think about doing a project, doing a group, doing community work, but something where we take psychoanalytic ideas out into the world.

LL: Yes. I think Kim's idea is a wonderful idea. I don't know how it will be actualized because psychoanalytic institutes tend not to be the most progressive organizations. But again, I came back from this recent Division 39 conference somewhat hopeful, because—almost all of the sections and committees in the division have become oriented to social justice. And sections have been changing their mission statements to be more active outside the clinic and more progressive. Section I recently tried to change its mission to focus on advocating for psychoanalysis in the public sphere, to challenge neoliberal thinking and the way "evidence-based" has been used to marginalize psychoanalysis and deprive underserved populations of adequate treatment. Section V changed its mission to "applied psychoanalysis."

EC: Fabulous.

LL: So things are happening.

EC: Yes, things are happening. Lynne, this has been wonderful, to have this chance to talk with you about the historical scope, depth, breadth of your work and thinking. It's been a complete delight and, as usual, I'm coming away with more to consider and more to explore. I look forward to seeing all the future contributions you'll be making to the field.

LL: Thank you. Thanks so much, Elizabeth. I look forward to more conversations.

EC: Thank you.

Bibliography

Aarseth, H., Layton, L., & Nielsen, H. B. (2016). Conflicts in the habitus. The emotional work of becoming modern urban middle-class. *Sociological Review, 64*, 148–165.

Alexander, M. (2010). *The new Jim Crow: Mass incarceration in the age of colorblindness*. New York: The New Press.

Altman, N. (2000). Black and white thinking. *Psychoanalytic Dialogues, 10*(4), 589–605.

Benjamin, J. (1988). *The bonds of love*. New York, NY: Pantheon.

Benjamin, J. (1990). An outline of intersubjectivity: The development of recognition. *Psychoanalytic Psychology, 7* (Suppl), 33–46.

Butler, J. (2005). *Giving an account of oneself*. New York: Fordham University Press.

Corpt, E.A. (2013). Peasant in the analyst's chair: Reflections, personal and otherwise, on class and the forming of an analytic identity. *International Journal of Psychoanalytic Self Psychology, 8*(1), 52–69.

Davoine, F. & Gaudillière, J-M. (2004). *History beyond trauma*. New York, NY: Other Press.

Fairfield, S., Layton, L., & Stack, C. (Eds.). (2002). *Bringing the plague: Toward a postmodern psychoanalysis*. New York, NY: Other Press.

Horkheimer, M. & Adorno, T.W. (1944/1972). *Dialectic of enlightenment*. New York: Continuum.

Jacoby, R. (1975). *Social amnesia*. Boston: Beacon Press.

Jacoby, R. (1983). *The repression of psychoanalysis*. New York: Basic Books.

Kohut, H. (1971). *The analysis of the self*. New York: International Universities Press.

Layton, L. (1998/2004). *Who's that girl? Who's that boy? Clinical practice meets postmodern gender theory*. Northvale, NJ: Jason Aronson. Reprinted by Analytic Press, 2004.

Layton, L. (2000a). Cultural hierarchies, splitting, and the dynamic unconscious. *Journal for the Psychoanalysis of Culture and Society, 5*, 65–71.

Layton, L. (2000b). The psychopolitics of bisexuality. *Studies in Gender and Sexuality, 1*, 41–60.

Layton, L. (2002a). Cultural hierarchies, splitting, and the heterosexist unconscious. In S. Fairfield, L. Layton, & C. Stack (Eds.), *Bringing the plague: Toward a postmodern psychoanalysis* (pp. 195–223). New York, NY: Other Press.

Layton, L. (2002b). Gendered subjects, gendered agents: Toward an integration of postmodern theory and relational analytic practice. In M. Dimen &

V. Goldner (Eds.), *Gender in psychoanalytic space: Between clinic and culture* (pp. 285–311). New York, NY: Other Press.

Layton, L. (2004a). Dreaming America/American dreams. *Psychoanalytic Dialogues, 14*, 233–254.

Layton, L. (2004b). A fork in the royal road: On "defining" the unconscious and its stakes for social theory. *Psychoanalysis, Culture & Society, 9*, 33–51.

Layton, L. (2004c). Relational no more: Defensive autonomy in middle-class women. In J. A. Winer & J. W. Anderson (Eds.), *Psychoanalysis and women (Vol. 32), The Annual of Psychoanalysis* (pp. 29–57). Hillsdale, NJ: The Analytic Press.

Layton, L. (2004d). That place gives me the heebie jeebies. *International Journal of Critical Psychology, 10*, 36–50.

Layton, L. (2004e). Working nine to nine: The new women of prime time. *Studies in Gender and Sexuality, 5*, 351–369.

Layton, L. (2005a). Beyond narcissism: Toward a negotiation model of gender identity. In E. L. K. Toronto, G. Ainslie, M. Donovan, M. Kelly, C. K. Kieffer, & N. McWilliams (Eds.), *Psychoanalytic reflections on a gender-free case: Into the Void* (pp. 227–242). London, UK: Routledge.

Layton, L. (2005b). Notes toward a nonconformist clinical practice. *Contemporary Psychoanalysis, 41*, 419–429.

Layton, L. (2006a). Attacks on linking: The unconscious pull to dissociate individuals from their social context. In L. Layton, N. C. Hollander, & S. Gutwill (Eds.), *Psychoanalysis, class and politics: Encounters in the clinical setting* (pp. 107–117). London, UK: Routledge.

Layton, L. (2006b). Racial identities, racial enactments, and normative unconscious processes. *Psychoanalytic Quarterly, 75*, 237–269.

Layton, L. (2006c). Retaliatory discourse: The politics of attack and withdrawal. *International Journal of Applied Psychoanalysis, 3*, 143–155.

Layton, L. (2006d). That place gives me the heebie jeebies. In L. Layton, N. C. Hollander, & S. Gutwill (Eds), *Psychoanalysis, class and politics: Encounters in the clinical setting* (pp. 51–64). London, UK: Routledge.

Layton, L. (2007). What psychoanalysis, culture and society mean to me. [Guest Editorial]. *Mens Sana Monographs, 5*, 136–147.

Layton, L. (2008a). Relational thinking: From culture to couch and couch to culture. In S. Clarke, H. Hahn, & P. Hoggett (Eds.), *Object relations and social relations: The implications of the relational turn in psychoanalysis* (pp. 1–24). London, UK: Karnac.

Layton, L. (2008b). What divides the subject? Psychoanalytic reflections on subjectivity, subjection and resistance. *Subjectivity, 22*, 60–72.

Layton, L. (2009). Who's responsible? Our mutual implication in each other's suffering. *Psychoanalytic Dialogues, 19*(2), 105–120.

Layton, L. (2010a). Irrational exuberance. *Subjectivity, 3*(3), 303–322.

Layton, L. (2010b). Maternal resistance. In J. Salberg (Ed.).), *Good enough endings: Breaks, interruptions and terminations from contemporary relational perspectives* (pp. 191–210). New York, NY: Routledge.

Layton, L. (2011a). The psychology and politics of privilege. *The Psychoanalytic Analyst.* Retrieved from http://www.apadivisions.org/division-39/publications/newsletters/activist/2011/04/politics-of-privilege.aspx

Layton, L. (2011b). Resistance to resistance. In A. Harris & S. Botticelli (Eds.), *First do no harm* (pp. 359–376). New York, NY: Routledge.

Layton, L. (2011c). Something to do with a girl named Marla Singer: Capitalism, narcissism, and therapeutic discourse in David Fincher's *Fight Club. Free Associations, 62*, 112–134.

Layton, L. (2013a). Dialectical constructivism in historical context: Expertise and the subject of late modernity. *Psychoanalytic Dialogues, 23*(3), 271–286.

Layton, L. (2013b). Psychoanalysis and politics: Historicizing subjectivity. *Mens Sana*, 68–81.

Layton, L. (2014a). Grandiosity, neoliberalism and neoconservatism. *Psychoanalytic Inquiry, 34*(5), 463–474.

Layton, L. (2014b). Maternally speaking: Mothers, daughters and the talking cure. In P. Bueskens (Ed.), *Mothering and psychoanalysis: Clinical, sociological and feminist perspectives* (pp. 161–176). Bradford, ON: Demeter Press.

Layton, L. (2014c). Some psychic effects of neoliberalism: Narcissism, disavowal, perversion. *Psychoanalysis, Culture & Society, 19*(2), 164–178.

Layton, L. (2015). Beyond sameness and difference: Normative unconscious process and our mutual implication in each other's suffering. In D. Goodman & M. Freeman (Eds.), *Psychology and the Other* (pp. 168–188). Oxford, UK: Oxford University Press.

Layton, L. (2016a). Commentary on Kernberg and Michels. *Journal of the American Psychoanalytic Association, 64*(3), 501–510.

Layton, L. (2016b). On moralism and ethics: Associations to Henry Abelove's "Freud, Male Homosexuality, and the Americans.". *Studies in Gender and Sexuality, 17*(2), 95–101.

Layton, L. (2016c). Yale or jail: Class struggles in neoliberal times. In D. M. Goodman & E. R. Severson (Eds.), *The Ethical Turn* (pp. 75–93). New York, NY: Routledge.

Layton, L. (2016d). What to teach? Social psychoanalysis in the clinic and the classroom. *Psychoanalysis, Culture & Society 21*(4), 422–425.

Layton, L. (2017). Racialized enactments and normative unconscious processes: Where haunted identities meet. In J. Salberg & S. Grand (Eds.), *Haunted dialogues: Conversing across history and difference*. New York, NY: Routledge.

Layton, L., & Schapiro, B. (Eds). (1986). Narcissism and the text: Studies in literature and the psychology of self. New York, NY: New York University Press.

Layton, L., Hollander, N. C., & Gutwill, S. (Eds.). (2006). *Psychoanalysis, class and politics: Encounters in the clinical setting*. London, UK: Routledge.

Layton, L. B. (2014). Normative unconscious processes. In T. Teo (Ed.), *Encyclopedia of critical psychology* (pp. 1262–1264). New York, NY: Springer.

Leary, K. (1997). Race, self-disclosure, and "forbidden talk." *Psychoanalytic Quarterly, 66*, 163–189.

Makari, G. (2008). *Revolution in mind*. New York: HarperCollins.

Mitchell, J. (1974). *Psychoanalysis and feminism*. London: Allen Lane.

Rizq, R. (2015). Perversion, neoliberalism and therapy: The audit culture in mental health services. *Psychoanalysis, Culture & Society, 19* (2), 209–218.

Ryan, J. (2014). "Class is in you": An exploration of some social class issues in psychotherapeutic work. In F. Lowe (Ed.), *Thinking space* (pp. 127–146). London: Karnac.

Silva, J. (2013). *Coming up short*. New York: Oxford University Press.

Stolorow, R.D. & Atwood, G.E. (1994). The myth of the isolated mind, Ch. 17. *Progress in Self Psychology, 10*, 233–250.

Wacquant, L. (2001). Deadly symbiosis: When ghetto and prison meet and mesh. *Punishment & Society, 3*(1), 95–133.

Elizabeth A. Corpt is past President, Supervising Analyst, and Faculty Member of the Massachusetts Institute for Psychoanalysis, Teaching Associate, Harvard Medical School, Department of Psychiatry at the Cambridge Health Alliance Program for Psychotherapy. She has written, published, and presented nationally and internationally on topics such as clinical generosity, the impact of social class on the forming of an analytic identity, and relational ethics.

The Psychic Life of the Political: A Conversation with Derek Hook
(Interviewed by Heather Macdonald, Lesley University)

Derek Hook began his career in South Africa as a lecturer in psychology at the University of Witwatersrand in 1999. It was there he developed an interest in the work of Michel Foucault and the analytics of power. Major themes in his scholarship have included developing discursive approaches to subjectivity as well as methods for critical discourse analysis.

In 2004, he took up a post at the London School of Economics where he lectured in social psychology until 2011. This was followed by three years in the Department of Psychosocial Studies at Birkbeck College. Between 2007 and 2013, Derek was a Dr. Hook was a trainee in the psychoanalytic training program of the Center for Freudian Analysis and Research in London. He currently holds an associate professor position at Duquesne University.

Dr. Derek Hook is also a practitioner of psychoanalysis with expertise in the area of critical psychology and psychosocial studies. His most

D. Hook (✉)
Duquesne University, Pittsburgh, PA, USA
e-mail: hookd@duq.edu

H. Macdonald et al. (eds.), *Dialogues at the Edge of American Psychological Discourse*, Palgrave Studies in the Theory and History of Psychology, DOI 10.1057/978-1-137-59096-1_10

recent research interests converge on the theme of the psychic life of power and his publications tend to either take up psychoanalytic, post-colonial or discourse analytic perspectives on facets of post-apartheid South Africa.

His works and the contributions to the field have been prolific with numerous books and articles. His edited book volumes include *Self, Community and Psychology* (2004), *Critical Psychology* (2004), *The Social Psychology of Communication* (2011), *Race, Memory and the Apartheid Archive* (2013), *Voices of Liberation: Steve Biko* (2014). He also has numerous monographs, *Foucault, Psychology and the Analytics of Power* (2007); *A Critical Psychology of the Postcolonial: The Mind of Apartheid* (2011); *and A Post Apartheid Conditions: Psychoanalysis and Social Formations* that just came out in 2013.

Heather Macdonald (HM): To give the audience a bit of background, perhaps we can start with some of your story. What drew you to the field of psychology in the first place, and what influenced your entry into the discipline and who shaped your initial ideas along the way? And I'm particularly interested in how you arrived at Foucault and not Marx or Marcuse, for example, as– because many people and many of your colleagues in South Africa certainly took up Marx instead of Foucault.

Derek Hook (DH): I think perhaps the pivotal moment, really, was Foucault. It's odd remembering some of these things—the experience of being interviewed gives you a - an objectification of yourself you may not have had otherwise.

HM: Yes.

DH: So I suppose what really happened was there were a bunch of us, colleagues who were studying psychology (Hook, 2004a;

Terre Blanche, Bhavnani and Hook, 1999). And this was now in the mid to late 1990s in South Africa. The apartheid regime was falling apart and there was very much a—a transformation of values. Mandela was out of prison. And the only thing that really seemed to be able to describe or say something about reality to me, and many of my colleagues who were doing psychology at the time, was Foucault, and more particularly Foucault's notion of discourse (Hook, 2001b, 2001c; Hook and Harris, 2000).

What was happening in South Africa at the time was a bit like two tectonic plates, two discourses chafing together like that. And so Foucault opened things up. A lot of us had also been reading a Freud and there'd been a strong psychoanalytic tradition but the answer to your question really is, certainly for me, it had to be Foucault. I was aiming to become a psychotherapist, and this early point in my career was also a gestation point for a series of subsequent critical reconceptualizations of psychotherapy as modality of power (2001d, 2003a, 2003b, 2003d, 2004c, 2004d). This work really brought Foucault into the picture, and the more Foucault I studied, the more it became apparent that micro-spheres—such as that of the therapeutic—could not be adequately analyzed without using critical tools that necessitated an analysis of broader spheres of power. Foucault was a crucial articulation point in opening the psychological to a broader series of political conceptualizations; he was absolutely central in politicizing the insulated and depoliticized field of psychology. You could write an interesting book on the forms of ostensibly critical theory and philosophy that were circulating in the South African academy toward the end of apartheid which should have been far more radical, which should have had more critical impetus than they did.

To give one example—and today the critique is made pretty strongly—there were lots of forms of Marxism, which didn't seem to quite get it with the racial element.

HM: Yes.

DH: In other words—and Fanon had long since anticipated this—you could imagine any number of pretty strident Marxists who somehow didn't really focus on the over-riding importance of *racial* dynamics, or perhaps didn't want to (Hook, 2001a, 2004a). And similarly, another case in point, was phenomenology. It seemed odd to me, that, certainly in Grahamstown in the Eastern Cape, you had this flourishing tradition of phenomenological psychology which, didn't really threaten apartheid values. So, for me, it had to be Foucault . . . Or more crucially, Foucault's (1977) sense of how a discourse is a violent thing, and his sense of, the historical emergence of something new.

HM: Yes, and it also seems like Foucault, his idea of power, it's more like an oil spill. So, power seeps into everything and it follows all of the appropriate gradients, and that seemed to speak to the apartheid condition as well.

DH: Yes. I should have said that as well. I just see discourse and power as omnipresently linked within Foucault. Another little phrase, which I would only become familiar with because I think Judith Butler only publishes the book in 1997 if I'm right . . . the "Psychic Life of Power—"

HM: Yes.

DH: Not necessarily in how she theorizes it, but that phrase became absolutely crucial in thinking about pretty much anything within the domain of psychology and politics for me. So, yeah, it starts with Foucault and although there was a lot of psychoanalysis in the background (Hook 2003a, 2003b, 2004c), it seemed to me that you need some way of engaging both with the discursive mechanisms of power (Bowman and Hook, 2011; Hook, 2005c, 2005e, 2013a) and the unconscious of power (Hook, 2006, 2008a), how these came to be replicated in fundamentally racializing ways (Hook, 2005a, 2005g, 2006, 2008e).

HM: So, shifting gears a bit, how has the field of psychology responded to your work? And we've talked before that you don't feel like there's been much response there because you've actually been working from the outside of psychology?

DH: It's a tricky question to answer because I don't think there's been much response within mainstream psychology. But one of the opportunities afforded me by this interview is also to think about what kind of psychology I've inhabited. And certainly in South African part of the changes that I was referring to, what was exciting about that is you could redefine the contents and the resources that you would use for a post-liberation psychology (Hook, 2004a, 2008d; Ratele and Hook, 2004). I'll say more about that later. But I suppose what it means is I don't think I've ever really done all that much mainstream psychology.

HM: You have had to mostly work outside of psychology to accurately construct a critique of the discourse.

DH: Yeah. So, I mean, I suppose what I—I had a little bit of a startling experience in London because at the London School of Economics, they wanted me to focus very much on the sub-discipline of social psychology (Hook, Franks and Bauer, 2011) rather than range between psychoanalysis, critical psychology and cultural studies (Hook and Howarth, 2005; Parker and Hook, 2008). The movement across disciplinary boundaries in South Africa seemed a little bit easier (Hook, 2001a; Hook and Parker, 2002). Certainly when I ended up in the discipline of Psychosocial Studies at Birkbeck College in London, this movement between disciplines was much easier, and in fact encouraged. In Psychosocial Studies they promoted and encouraged the move to use certain forms of social theory in response to ostensibly psychological probelmatics (Hook, 2008b) . . .

HM: I see. It also seems that one of your driving—or important ideas —is the relationship between the psychic and the social. And so you seem to get sometimes, like you're describing in London, forced into one corner or another. Disciplinary discourse itself is divided where one is looking at the components of psyche, or at larger social/global processes. However, your work suggests that these domains are not separate at all and inform one another on all levels.

DH: Yes.

HM: That's the kind of psychology you're doing, a psychopolitical analysis of power, and oppression, not only in their imaginary and symbolic forms but also how these forms are embodied.

DH: It is curious to me. I don't know yet. I haven't been in the States for very long, but I'm guessing that the notion of Psychosocial Studies as a standalone discipline doesn't really exist, or people don't have a good sense of what it is. Be that as it may, that was something, which was of crucial importance to me. And, interestingly, just reflecting on some of these questions for today, I remember being—the only other time I've been in Boston was in 2002, and I was visiting with a colleague. And she took me to a radical bookstore somewhere in Boston. I don't know, maybe some people know where it is, if it's still there.

And up on the wall, there was a whole bunch of postcards. And one of them was a Steve Biko postcard and it said something along—the quote was, "The most powerful weapon in the hands of the oppressed is the mind of the oppressor."

HM: Right.

DH: I mean, this thing just drew me to it. Number one because Steve Biko's, obviously got an illustrious history within South Africa. For those of you who don't know, Steve Biko was an anti-apartheid leader who was killed by South African security forces in 1977.

HM: One of the most brutal assassinations in history.

DH: Yes, and he was an inspirational leader of the Black Consciousness Movement in South Africa. So it was nice to see this thing in Boston. But also, the quote was just perfect. Inasmuch as for me, that's—that's the focus, right?

HM: Right.

DH: It's not—it's not psychology given in terms of its own terms of whatever, intelligence, personality, all those things. It necessary must be psychology—or the psychic—in its relationship to the social and political domain. So, the Biko postcard was important in a number of ways (Hook, 2011d, 2014b). But in fact, this helps me respond to your earlier question. That's my discipline, really. My discipline's not psychology in the more limited sense, it—it's closer to a discipline that studies *the psychic life of power*. And that's, I think, in some ways, how colleagues in London would think about psychosocial studies. It does that. And in fact, just to add one further point—one you have already made—maybe it's not just simply the psychosocial, it's also should be the psycho *political*.

HM: Yes. Well, and the less intuitive idea that power itself has psychological facets, and I think that is more subtle—you reverse the equation a little bit and say power is constructed psychologically.

DH: That's right.

HM: And with psychoanalysis as your ally you are able to link fantasmic or libidinal investments with the social constructions of space, identity, and political ideologies (Hook, 2011e, 2012a, 2012b, 2015).

DH: Yes.

HM: Speaking of Biko, in your book *A Critical Psychology of the Postcolonial* (2011), you rely heavily on Biko and Fanon as thinkers who engage the psychopolitical using multiple registers. But these are also figures who've appeared to transcend and transform the social imaginary and the symbolic by going beyond it. How do you account for this type of transformation, both at the individual and social levels? Is this, and this is always my question, is this type of genius available to us in the trans subjective social structures? And if so, then what would be the political model for revolutionary practice within the field of psychology?

I realize this is not a small question.

DH: Let's think about that. I mean, maybe to pick it up in a narrative way and say, "Okay, we started off a little bit with Foucault, what happened then?" I suppose after Foucault it was Fanon (Hook, 2005c, 2008d, 2011a; Hook and Truscott, 2013). Fanon spoke to the South African context equally powerfully, if not more powerfully, and that opened up that psychosocial, psychopolitical dimension as well. I mean, that is—that's also omnipresent, I think, in his work. And I think it's—for me, it's a love story. I love what he does critically, politically, intellectually because you have all these wars within Fanon's studies. Some people say, "No, don't get all psychoanalytical." He's not that, he's not just a psychoanalyst.

Then you get other people who say, "No, actually you have to read *The Wretched of the Earth* (1963/1986), and you must just see him in that sense as a revolutionary." I think you need to keep both of those pieces of Fanon together, and you need to run them together (Hook, 2011a). And ultimately, that feels like a bit of an impossible thing to do because it's very difficult, if not impossible, to do both of those types of analytical work at the same time; the psychic life of power as well as the structural life of power. And they seem to separate and merge and come back and come forth.

I think—I'm not sure Fanon's always able to keep them both in the same picture. There's the problem of, "Well, is that psychological reductionism, or is that sociological reductionism?" But he keeps on trying. So that—that's a crucial intellectual high point for me.

Back, though, to your question, what else draws me to Fanon and how else does he help think a revolutionary agenda? I mean, I'm not a revolutionary, and I know that now. Interestingly, Biko was the one person, I think, who made me realize that. So Biko's got this longstanding critique of white liberals (Biko, 1978), particularly in South Africa (Hook, 2011d).

HM: Yes.

DH: I could smuggle that in here because it's a nice point to make. And basically he (1978) says, "You get lots of left wing white South Africans who make a lot of noise and make a lot of like, "Hey, I'm liberal, I'm leftist," or whatever." And he says, "Ultimately, these guys are worse for the struggle in South African than hard-line, right wingers who want segregation," because they try to win on two fronts. They want to have a good conscience by making a little penance and show that they're doing something good. But they don't ever do anything and they're not really actually ever going to sacrifice anything.

So, I realized at some level, are you playing that game a little bit? Anyways, that's perhaps another conversation.

So I think I had to make peace with myself, I'm not as radical as I would like to have thought I was. And I'm not a revolutionary.

HM: I think Biko would agree . . .

DH: Yeah, well—I mean, which is not to then say, "Well, you shouldn't be doing anything political," right? Okay, so back to Fanon. What I think is so enabling and so potent in Fanon, which is both a political and an intellectual and indeed, in some respects, an aesthetic objective, is that it enacts seven or eight different genres simultaneously. So *Black Skin, White Mask* (1963/1967), is composed of little fragments of novels, of psychiatric bits. You see that in *The Wretched of the Earth* (1963/1986), as well, little psychiatric and indeed psychoanalytic case studies at the end. There's a form of phenomenological existentialism. There's reference to various different psychoanalysts and psychoanalytic, psychological thinkers.

It is this exercise in *bricolage* where he puts all these things together. And for me, what was useful about that is—well, it was—it was disturbing to read Fanon because I couldn't understand what was going on. I don't know if I've ever quite managed that. And, I mean, you must have all had this. Sometimes you get drawn to a book. A, you don't understand it; and B, it disturbs you. And C, there's the signature image or problem.

For Fanon, there's this—he gives this famous example of a scene on a train where we hear a child saying to their mother "Mommy, mommy, I'm afraid. I see a Negro." And then he describes this bodily experience of breaking apart of being burnt—a whole series of images of black blood splattering, black bodies being eviscerated. And I used to try and figure out what was going on with this (Hook, 2013d). And I thought, "No, it's okay. It's a poetic metaphor." And then it dawned on me, no, it isn't a poetic extended metaphor. And there's obviously a history of lynching and all this kind of thing which means that it's not solely—it's not solely poetic.

But it just—I never quite managed to digest it. I started writing other stuff after writing on Foucault. And this—part of this challenge of trying to comprehend something in Fanon kept on coming back. And maybe we'll talk a little bit more about some research that linked from that. But that's problematic, that traumatic, bodily destructive image that's in the text is also reflected in the form of the text. Because the book itself seems to be cut up and reassembled, like a body in pieces, to use the Lacanian phrase.

And for me, part of what was so inspiring about that is he can start to do analytical things that you would never be able to do if he was a diehard Freudian or a purist Marxist. And you see him wrestling with, and sometimes not always successfully, innovating a whole series of new concepts. So, he starts talking about, and I think in an incredibly innovative way, the racial gaze. He draws on the Freudian notion of scopophilia, the drive to look, but it's also about power, it's also about racializing embodiment (Hook, 2008e, 2011a), how the power of the gaze not simply objectifies but racializes.

There's numerous other examples that I could cite, but for me that's the real potency of what he does. He gives you a novel critical register where no previous critical registers had done the job quite sufficiently. So he takes to task Octave Mannoni, the guy who does all these other kind of apparently psychoanalytic analyses—analyses of colonial power. And he rips them apart and he does something that's still somehow indebted to it, but nonetheless different.

I think maybe that's only a modest way of trying to get around to answering your question about transforming the social imaginary. And he was a revolutionary, right? Okay, but at the intellectual level, at least, that ability to come up with new critical tools with which to cut, with which to critique, is something that's formidable, it is for me. And that would be the closest, I think, I can come to answering that question.

HM: Yes. He also engages clinical diagnostics at the same time. He sees apartheid power, for example, as phobic, and this is his diagnosis of apartheid power. So he engages diagnostics along with the rigorous critique that you're discussing.

DH: Yes, and there again is the intersection of the psychic and the political. Something else occurs to me here, in response to the influence Fanon has had on me. I noted above Fanon's reference to the broken, destroyed body. The question of the body and embodiment had been of interest to me for a long time, because it seemed to represent a domain somehow—at least in part—outside of the immediate domain of discourse (Hook, 2002, 2003c, 2003d, 2008e). Fanon's account of the experience of inhabiting a broken, destroyed, or disrupted body was striking to me. When I seriously started considering a type of psychical analysis of post-apartheid South Africa (Hook, 2012a, 2012b, 2012c, 2015) and recalling my own memories of being a kid in apartheid South Africa, this motif kept on coming back. I am speaking about the prevalence of this image of a broken body. And not just anyone's broken body, a burned body, a destroyed body, a body blown to pieces—it was a *black* body (Hook, 2013d).

So this seemed to me both disturbing, but also needing some kind of analysis. So I started trying to think about how do you go about doing this research project, of these images? And I had a series of personal memories, and one of which was being nine or ten and it being a Sunday morning and going off to get the newspaper for my parents. And on the front page of the newspaper, there was a picture of an obliterated body. There was a picture of an Umkhonto we Sizwe operative—an operative that is of the armed wing of the African National Congress. A guy tried to lay a bomb and the bomb had blown up and it destroyed him. But it was on the front page of the newspaper and the back page, like this kind of gate-folded double page spread.

Now, I was young when I saw this. I mean, I wasn't going to do an analysis of the political dynamics of representation, but it just—it struck me like something's wrong with this. I don't—I don't, I didn't really have the words. And when I started to write about the theme of the body-in-pieces, I realized that I wanted to involve that anecdote, I kind of had to. It was almost as if the topic was pulling me in and was also hooked onto my own fantasy because I realized that I was still somehow fascinated. I hadn't been able to get rid of that image.

Then I found multiple different versions of it . . . A year ago, I was visiting my extended family in South Africa and they have a little community newspaper, throw-away newspaper. And just in the middle of page three, there's a picture of a body next to a railway. It's a black body. And you read it and wonder: why is it there? And there'll be some caption: "Man found deceased on the side of the railway." But it bothered me, why even include the picture? Why take the picture? And why put it in the community newspaper?

Nevertheless, this image I found in lots of different places in South African popular culture, even in post-apartheid South Africa where you would have thought things would change. Or that you may also start to find white bodies, although you don't.

So I started writing about—I did a presentation. I did a presentation in New York about it where I was attacked. People said to me "What's going on here? Don't you realize that these are real people?" And I said, "That's the whole point. They *are* real people . . . " What's disturbing to me is that clearly there's this fantasy, this fantasmatic preoccupation with the black body, or the black broken body, which still exists in South Africa.

I suppose the charge was: you're still recycling racism by even showing the images. Your interest in the images is somehow problematic inasmuch as it extends your own fantasy. That was a telling experience for me because it meant that I was personally implicated in the research that I was doing.

One could then ask: at what level was this research happening? On the one level, you could say you could do a whole dynamics of formal representational codes that are being utilized to make this thing shown, to make it a depicted image that has this currency—as an object of fascination—that it still seems to possess. But clearly, I was also engaging with it at a personal level. I suppose what I'm saying is that in the psychosocial work that I'm doing, you are also totally a part of what you're researching, and part of your own psychic investment is going to become apparent in what you're researching. And ultimately, I think it was beneficial to me. It wasn't a nice experience to be accused of racism like that, but it—it was necessary, to see the methodological and political—psychopolitical—problems. It wasn't simply a question of how do you gauge with the materiality of individual text or image, but of considering also *how implicated you are in the thing that you find interesting* in the first place.

HM: One question, just to follow up on Fanon, he concludes—his conclusion is violence. And I'm wondering your thoughts on that? Because he suggested there was no escaping violence in both the colonial situation and the postcolonial situation.

DH: I suppose there are different things to say about that and there are different conclusions. I mean, the violence thing is obviously in *The Wretched of the Earth* (1963/1986) rather than *Black Skin White Mask* (1963/1967). So, in *Black Skin White Mask*, there is an explicitly humanistic response to the problems of colonial racism. Then again, while the humanistic hope of *Black Skin White Masks* seems very different to the recourse to violence apparently advocated in *The Wretched of the Earth*, one should recall that ultimately this is—odd as it may sound—a humanistic appeal to violence. That is, the appeal to violence in *The Wretched of the Earth* is not a violation of humanistic principles or ideas, but precisely in the name of humanity, so as to secure the status of the human for those subjugated, objectified, designated as less than human. So I would suggest this qualification in respect of Fanon's humanistic appeal to violence, it is first and foremost an appeal for the humanity of those damned to what he elsewhere refers to as the "zone of nonbeing," the dehumanizing objectifications of colonial racism.

The difficulties of successfully responding to the petrification of racism should not be under-estimated. In *Black Skin White Masks*, Fanon makes reference to the idea that both the white and the black subject are, respectively, locked into their whiteness, their blackness. This Manichean impasse is seemingly irresolveable within the colonial context, and cannot be rectified merely by reference to idealized images (fantasies) of racial harmony, togetherness, sentimental ideas, as in South Africa of a "rainbow nation." This lockedness is both bodily and discursive.

HM: This is why the idea of racialized embodiment is so important.

DH: Yes. That speaks volumes about how, in post-apartheid South Africa, one ends up being insulated in one's racial shell.

HM: Yes. Back to the topic of violence.

DH: One obvious and pertinent political message here is that the way one wins freedom, the way one wins liberty, has to be violent. By which one means you can't have it in the way that many would argue did happen in South Africa where liberty comes as a kind of charitable endowment.

HM: Or through a truth and reconciliation process.

DH: Yeah. Or, it comes as a gift or a, "Okay, here, you can have your liberty." And for Fanon, that doesn't work for very obvious reasons, because it still means that who was in power remains powerful through the symbolic act of saying, "Okay." Which also links back to a whole series of Steve Biko's (1978) arguments about the problems of certain forms of white anti-racism that adopt the form of charity (Hook, 2011d).

HM: Good, yes. I want to move into the theme of repetition.

DH: Sure.

HM: Because this is an important theme in your work. It's also an important theme in many of the discourse that's coming out of South African right now. This notion of history constantly repeating itself even 20 years later after the end of apartheid (Hook, 2012a, 2013c, 2015). And Lacan uses Kierkegaard's distinction between repetition and recollection to help him articulate his own distinction between the symbolic, which would be repetition, and the imaginary, which he identifies as recollection. So thus, repetition for Lacan happens in search for the lost object, melancholia, and within the relation between desire and the object that is perceived to fulfill that desire.

So, however, for both Kierkegaard and Deleuze, what returns in repetition is not the same, but different. So, in many ways, we don't know what repeats in repetition. This is a point that Mbembe and others often make when they're critiquing structuralists such as Levi-Strauss or Lacan. So how do you see the notion of repetition playing out in the post-colony? Are Lacan's ideas of repetition useful or, more importantly, do they contribute to conceptualizations of temporality that add to the oppressive discourse in the post-colony?

DH: Okay, so the thing is with Lacan, a bit like Fanon and other people, I do also have a bit of a love affair with him. I am very proud of the work where I have used Lacanian ideas to further extend an analysis of power or move beyond the paramaters of social psychological analysis (Hook, 2009, 2010a, 2013a, 2013e, 2013f, 2014c, 2016a). But it's a tempestuous love affair inasmuch as I—I alienate myself with Lacan sometimes, trying to read him too much, too faithfully. And it becomes at some point unproductive inasmuch as the best thing about Lacan is that he's so innovative in his reading of Freud. And if one's trying to mime or be faithful, to Lacan, I think one should try and be innovative in terms of how one uses him. So, sometimes it's problematic to apply those concepts without some—without some added critical impetus (Hook, 2011e; Hook and Neill, 2008, 2010).

Having said that, that meta theoretical issue of temporality, repetition, repetition in its various guises, is revisited by him, and very many other people in different kinds of ways, as you've suggested. He has a terrific contribution to make in terms like automaton and tuché, that is, the traumatic event which is somehow linked to something that happened in the past as opposed to automatism's symbolic reiteration of something that already happened.

So leaving that as a backdrop, I think the most pertinent way to answer your question is that for me one of the most interesting and crucial issues in psychosocial studies, or if you want to call it critical psychology, or theoretical psychology's approach to sociality, is temporality, as you've suggested (Hook, 2011b, 2015).

HM: Yes.

DH: The dimension of space, or as it is sometimes referred to 'spatiality,' is clearly of great importance in studying power be it in contemporary South Africa (Hook and Vrdoljak, 2002) or elsewhere. I drew on psychoanalysis to undertake analyses of this sort in the (post)apartheid context, particularly in view of monumental space (Hook, 2004b, 2005d). Important as this was, the question of the political dimension of temporality had been somewhat neglected, especially from a psychoanalytic or psychosocial point of view. Temporality becomes particularly fascinating in post-transitional societies, and again South Africa's a case in point, where you see various forms of staggered, alongside various forms of retrogressive or reverse temporality (Hook, 2012c, 2013c). And you see so many multiple modes of discontinuous temporality happening at the same time. So, I mean, you've—you've just come back from South Africa and you will have probably had some of had that experience. That if you travel through the country, you will have moments where you experience something like a type of fast tracked temporality, right?

Where things are seen to be a little bit more in the future than you would expect for South Africa, somehow more ahead of other places in the world. And, of course, you also get the reverse situation where you feel like, "Whoa, I've just stepped out in some very different space and I feel like I'm in 1964." But you also get—and I think the work that I've been trying to do of late focuses on this—experiences of "petrified life" where you find a variety of forms of stasis or suspension. And so for one example, Achille Mbembe (2001), who you referred to, talks about repetition compulsion, as "repetition and difference." He takes it from Fanon who, of course, is using it from Freud and he talks about how—what it means to be oppressed is to be oppressed in one's experience to temporality as well. And that what it means, in certain forms of oppression, to be oppressed and to have an oppressed temporality, is to continue to repeat without being able to break out of that short circuiting cycle and be able to make a different future.

HM: Right.

DH: There's also an odd white nostalgia for apartheid (Hook, 2012c; Stevens, Duncan and Hook, 2013).

HM: Yes, a strange mixture of temporal realities.

DH: You see this odd sort of hysterical nostalgia for a time that was, and that now is being lost, in all sorts of different ways, including architectural formations. So, it's interesting then because even though you could say, "Well, of the various forms of social division and social discontinuity that separate various groups in South Africa, you would have thought that that would occur along temporal lines as well." I'm talking about suspended temporality, about different forms of racialized temporality. And indeed, you could argue that it does. But oddly enough, you could say that both oppressed and former oppressor, seem to experience a similar suspendedness of time, even if in different kinds of ways.

In some of the material that I was looking at to try and get a sense of suspended temporality—this, interestingly, was also a theme at the end of apartheid—I found the work of Vincent Crapanzano, the American anthropologist, who writes this book . . .

HM: Such as the 1985 book *Waiting*?

DH: Yes, it's a fascinating analysis where he says, "I spent a lot of time with the whites of South Africa in the eastern Cape." And he

> records a series of interviews and he tries to do this whole ethnography about how they're experiencing things. And for him, the single most pertinent motif from all of it is the idea of *waiting*, waiting for something to happen. But a dreadful waiting in which one is suspended and you don't know what's going to happen, and it becomes a waiting which ossifies time and objectifies things and deadens the potential openness and spontenity of human subjectivity.

My notion of petrified life (Hook, 2015) builds on that and tries to further articulate it, but also finds this other fascinating thing, that the Apartheid Archive project (Hook, 2011b, 2013c, 2015; Hook and Long, 2011; Stevens, Duncan and Hook, 2013), which some colleagues and I are involved with, it's a whole collection of narratives of apartheid experiences collected by us from a number of universities. It's not a very big archive, but some of the narratives we've collected about the experience of apartheid racism talk about time. And one was particularly interesting to me. It was about a lengthy discussion—not a discussion, a description—of a young boy, obviously a, whatever, a middle aged man writing, remembering things. And he describes the scene of finding a whole series of odd instruments around his parent's house. There's an ax, there's a hammer, —and there's something out of joint about this experience.

And, of course, it gradually becomes apparent that these household items are old weapons, or potential weapons. And as the story unfolds, he comes across a chest which has a whole series of old family items in it. Out he pulls a whole bunch of things including an assegai, a spear, a traditional Zulu spear. And then he finds a truncheon, a police truncheon. And the description goes on and he becomes sure in this moment—it sounds like a perfect instance of the short story genre, a moment of epiphany—that this truncheon has been violently used by his father on a black man.

Toward the end of the story, you get the answer to what's this pervading sense of dread and unease and congealed life. And it is that he thinks that without saying it, his entire family milieu is characterized by the unspoken sense that something will one day happen that will be the revenge, or the payment, that will be extracted from them as a family.

HM: For having done violence toward the black man?

DH: I mean, you never know that that violence actually happened. But anyways, the point of it all is—a kind of a Freudian thing—is that there's this guilt which becomes consuming. It's repressed. It doesn't get said as such, but there's an absolute, almost paranoid fearfulness about what changes are happening. And I think you've experienced some of that in South Africa as well, this white repressed discourse.

HM: Certainly, yes.

DH: So, that is another form of petrified life, and the ambiguity of this phrase comes through quite nicely because it is intensely fearful, as well as frozen. Because you don't want time to move too much further because inevitably at some point, there's going to be some sense of retribution. So I thought it was a rich theme. I don't know how many other empirical sources I can find containing a similar idea. But you can find it in something like J. M. Coetzee's novel *Disgrace* (1999).

HM: And I think Nadine Gordimer in *Burger's Daughter* (1980) has also written about that.

What you say really resonates because when I was in South Africa this past summer, I was staying with a white family in Pretoria. I mistakenly opened a closet thinking it was the restroom and in the closet were these bags, and they were all packed. I asked one of the family members, I said, "Well, why do you have packed bags in your closet?" They said, "Well, in case we need to leave. In case something happens." So I thought, "What do you mean, in case something happens? What's going to happen?" But nobody knows. And when you—when you ask either white or black South Africans what's going to happen in this waiting, in this petrified stasis, there's no real answer. I think that creates a lot of anxiety and aggression.

DH: In trying to get a sense of what is happening in a social milieu or the power dynamics of a given place, one should always ask about time, ask people about their experience of temporality. You do often get fascinating responses and thoughts that wouldn't have come out in quite the same way if you'd asked about history or whatever.

HM: Well, I think it's an important one because I can't remember who was it, perhaps Achille Mbembe or somebody else, who said that trauma is irreparable time. So that is—that's true, I think, of the South African context where you have time that cannot be repaired. And so you're in that constant limbo of can't go back to the past, certainly, and you can't move forward.

DH: Yeah.

HM: So, here is a short, but difficult, question. Is the goal to have a raceless or non-racial future, especially for South Africa?

DH: Okay, that is the most difficult question anyone's ever asked me, ever. [laughter] You get different responses in the literature. I used to read a lot of Paul Gilroy, the British sociologist and, of course, he's got a very strong argument about, moving beyond race, about what comes after race, and one should reach toward this. And you can anticipate the criticisms that follow from that type of argument—that this is naïve, that it's not yet time to think this way, etc. One needs to work through what race means and various investments in race.

So, here's one answer. I mentioned a couple of these people that I've been interested in reading and working with and on, Biko, Fanon, Foucault, maybe Lacan, too. But the other guy that I'm interested in at the moment is Robert Sobukwe (Hook, 2014b, 2016b), who nobody would have heard of presumably because he has a tragic history. In the late 1950s in South Africa, the African National Congress, South Africa's governing party, the party of Mandela, was confronted with a more radical rival: the Pan-Africanist Congress. The Pan-Africanist Congress was itself an off-shoot, from the ANC, from the mother organization, who wanted to do a properly Pan-Africanist thing. So you could see from the word go that this is a more visionary, more potentially revolutionary agenda. They wanted a strong vision of an entire united Africa.

Anyway, so Sobukwe emerges at the same time as Mandela and for a short period of time across the whole of Africa, many people think he's got far more promise as a leader than Mandela himself. He's imprisoned before Mandela, as well, and the more I read the literature, the more I feel that Mandela learns from Sobukwe. It doesn't end well for Sobukwe. The PAC collapses largely when he's in prison . . .

I have been trying to read a lot of Sobukwe's Pan-African Congress writings, and his answer to your question would be the Pan-Africanist Congress response, which would be to say: there is only one race, the human race. And then, of course, the obvious contradiction comes in. But you're calling yourself *Pan-Africanist* and believe and feel that people should identify very powerfully with Africa, white or black. And Sobukwe's response would be, "Yes, I do believe that." And you say, "But that's a contradiction, right? You're saying there's only one race, the human race. But you're also saying that you feel we should have a particular investment if we're *Africans* in the future and progress of Africa."

And it does sound like a contradiction. And I think that's the way lots of intellectuals who've engaged with it get stuck at that point and say, "Okay, it doesn't work. You can't pursue both of these things." And maybe they're right. But I think the way Sobukwe wants to try and deal with it is to say, "No, we've got a continental vision of Africa. But the reason that we motivate for Africa and for African people is because—not because we're still hung up on race, but because we're hung up on the issue of oppression. And we are focused on whoever is the most oppressed in a given situation; economically, culturally, whatever."

The dimension of temporality here is also important—because Sobukwe's not insisting that overnight that blacks and whites alike are simply all African. He's not saying, "I command—wave my magic wand and make us all African." Which incidentally, is an idea many whites like that because they can say, "I'm African, too."

HM: Yes.

DH: So he's not simply saying that, like the click of the fingers, "Now we're all African." This, certainly for many whites, in a future political destination, might be a goal that can be achieved when the structures of white supremacy have fallen. He's saying that there's a great deal to work for and a great deal of political work and a great deal of justice that needs to be dispensed and thought through. But he's also wanting to say, "We'll move toward this goal, there is only one race, race as a concept should fall by the wayside, but those identities that have been oppressed and that enable us to progress the fight against oppression—precisely such as African—need to be retained and developed until the point of practical structural equality is reached. This whole argument depends on a future goal-point; what he speaks of—certainly in terms of whites genuinely becoming Africans—is dependent on political labors in the present, it is not something that immediately comes into being. What we now need to do is the hard work of trying to figure out how best to respond to a whole series of systematic inequalities, injustices, which also, incidentally, means that it's not enough to do what the ANC does and say, "Let's have a broad front of a whole series of different political constituencies that we'll bring together in the Freedom Charter." He wants to say, "Well, let's consider who is most absolutely marginalized?"

Which also means, although he doesn't thematize it like this, that a latent form of feminism must be part of the political agenda, right? Because you become sensitized to the different dimensions of oppression and you start with that group which you feel is most—or that you can show to be most oppressed in certain ways. So that would be an attempt at answering an impossible question: there are two answers, one for now ("It is too soon just yet, until social justice issues have been attended to, for us to be beyond race") and another for the future ("When the structures of white supremacy have been destroyed, then whites can meaningfully be thought of as Africans, and the ontological condition of race—there being only one race—will be shown to be true").

HM: Yes, it is an impossible question. But I think the idea of working from the perspective of justice versus injustice along the lines of what Sobukwe's message—makes sense because this would also include economic injustices.

DH: Yes.

HM: Why—this is a little bit of an aside—but why do you think Sobukwe has been written so much out of the history books?

DH: There's lots of fascinating reasons. One of the most interesting is that there seems to be a kind of repression that is coming undone. I saw a colleague at Duquesne the other day who was doing a talk, "Gandhi as Metaphor," and it makes me think that I should do a talk that says, "Sobukwe as Metaphor." So Sobukwe is now starting to come back. Why was he been written out of history?

HM: Has he been repressed so to speak?

DH: Well, you could say there was a consensus of forgetting. Lots of white South Africans don't want to remember Sobukwe because they identified him as being a proponent of "throw all the whites into the sea" because of his strong Africanist agenda. I don't think that's what Sobukwe wanted.

HM: No.

DH: But then again, he—he did also realize the importance of a radicalizing discourse of that sort (allegedly "anti-white" discourse). So we could have lots of discussions about the various forms and representations of Sobukwe. He's not a proponent of violence, but he also does realize that you need some spark to get the political momentum of a certain political discourse going. So, white South Africa is quite happy to forget Sobukwe. He also says really critical things about Mandela.

HM: Yes.

DH: Before Mandela's even imprisoned. So this is, a different era that doesn't play so well today in an era where Mandela is, justifiably in some respects, lionized. But also, you could say that when the transition came, Sobukwe was on the losing side. I mean, he was dead by then, but the PAC became marginalized and were easily painted as radicals. And they—they didn't seem to have the nice harmonious reconciliatory narrative that the ANC had through Mandela.

So there's two reasons. There's lots of other interesting reasons, but for those of you who've been to South Africa, or if you go to Robben Island you hear a lot about Mandela and the ANC. And there's tiny little house where he was—Sobukwe was—imprisoned, for six years in solitary. The bus chugs past it and a lot of people have had the experience I've had where the tour guide says "Well, that's Sobukwe's house? Anyone want to stop? Let's carry on."

And it's a remarkable performative example of—I mean, the guy was a great hero—of being written out of history.

HM: Completely.

DH: Increasingly today there's a sense that the happy Mandela reconciliation vision sidestepped a whole series of issues of justice, or injustice. Sobukwe had it different. And here is another answer to your question: Sobukwe never, ever wanted to give up on land. Land was part of what liberation meant, it was part of what African identity was, part of a Pan-Africanist idea.

HM: Also, the decolonization of governments who would redistribute the land.

DH: Yes, this helps us understand the psychology of liberation. For Sobukwe, it's never merely psychology, it's always necessarily material and land is an issue. So you can have various forms of transition in South Africa, but despite many attempts at reconciliation, the land is not returned. And that's what Sobukwe felt was absolutely necessary. And that's also what freaks white people out in South Africa who own the land. They don't want to be told they're going to lose the land.

HM: Yes, it does. Yes, that's right.

DH: That's one of the reasons why I think he's still a radical and disconcerting figure in South Africa today.

HM: So, while we're on Sobukwe and Biko and Fanon, much of your work has been focused on what you term as these master signifiers. So Biko or Sobukwe, and I'm—I'm interested because you write so much about the psychoanalytics of power and racism and oppression and discourse and how these libidinal economies really play into that. So I'm wondering if there's overlap between these master signifiers of Biko and Sobukwe as well as—I mean, do they also operate in those same libidinal economies as racism and oppression?

DH: Ok, so let's come up with a methodological model and let's call it something that already exists, "libidinal economy", and try and say what makes a group or a discourse function at a given time with the vibrance of a certain affective field. And so the idea, in thinking through the prospects of such a methodological model, was that you would need to pinpoint, five or six or seven crucial questions. What does that group or discourse take as crucial to its survival? What does it cherish; what does it consider to be most fearful or threatening? What is it that enables the group or discourse to maintain its sense of psychic equilibrium?

So you can apply a whole series of similar questions that point to the libidinal underpinnings of the discourse or group in question. In doing this, you can involve a whole series of theoretical ideas. Slavoj Zizek (1997), for example, has this idea of racism as theft of enjoyment; Freud has the notion of the libidinal constitution of the mass which points us to ask: "What are the bonds, the ideals, the ego ideals, that hold the mass together?" We might return here to Fanon and the obsessionality of race. Fanon also makes a contribution to libidinal economy—he says that one way of thinking about racism is to consider how it functions as a phobia formation.

So I started trying to do that in post-apartheid South Africa, and indeed in post-Mandela South Africa, which is also a slightly different place. And many things seemed to crystallize around these big names, and what I started to realize was that Sobukwe, or Biko, or Mandela, are themselves effective formations. And here we can return to the idea of Sobukwe as metaphor.

So to talk about Biko is not simply, and sometimes not at all, actually, to talk about the singular "actual" historical subject. It's to talk about what he's seen to represent for a cross section of people and what he has come to mean today. So in that respect, my concern with the master signifier—as in: "What does Mandela mean today?"—it's partially about Mandela, but it's actually far more about how his name has entered into popular discourse, into a field of profound personal subjective investments in what he is imagined to mean, to stand for.

It's about tracking those formations of affect, formations of hope or fear or anxiety, or perhaps something else altogether. And so that's what I was trying to do in thinking about the names. And also, actually, when you try to do interviews, I'm not a very skilled interviewer, or ethnographer, or anything, but I tried. If you ask someone "What's your libidinal economy?", it's never going to work, right? But if you start to ask, "What does Biko mean to you? What does Sobukwe mean to you," you're immediately able to access some stuff. So that's, yeah. That's part of it.

HM: So, when you view the field of psychology from your perspective and from the kind of psychology that you've been doing, what are your greatest concerns about the field at large, and what would you think it would need to do to right its course? If you perceive that it's off course in some manner of speaking?

DH: That's also difficult because I suppose I don't really know what the field is anymore. One way of doing that is just to look and see what's in textbooks. One of the trajectories in my earlier career was that of critical psychology. And one part of critical psychology is the ideological critique of psychology. And my colleague, Ian Parker, does a lot of that. For my part though, I don't want to be the diagnostician of psychology.

I'm saying a lot of things which makes it sound like I position myself completely outside of psychology. But I realize that I don't do that, actually, that I've still got a significant investment in psychology. But my tactic, and it certainly was in my *Critical Psychology of the Postcolonial*

(2011), is to say that there's a whole series of probelmatics and dilemmas and concepts in psychology which remain profoundly interesting.

My approach is to read a whole bunch of different texts that would never be considered psychological, in order to ask how they inform a more explorative, mor expansive and critical psychological sensibility. So in *Critical Psychology of the Postcolonial*, I ask: "What if you read the South African author J.M. Coetzee? What does he have to tell you about the desire of apartheid ?" "If you read the South African biographer Chabani Manganyi, what does he tell you about psychology in Southern Africa?" And you can do the same thing with Biko, you can do the same thing with Fanon.

That was what was great about a certain period in South African critical psychology—you could basically redesign your syllabus. You could say, "First you're reading the psychology of Fanon, then we need to try and understand something about a psychology of resistance or psychology of anti-racism. And our reading today will be Biko." So you could do that. And I suppose that's still what inspires me, and that would be my suggestion about how to deal with the psychology: read lots of stuff outside of psychology.

HM: As far as your—the future of your scholarship, would you be able to give us a sneak peak of—of your future and upcoming works?

DH: I think I'm going to retire. No, not really. [laughter] No, I suppose one thing—I don't know whether I'm able to do this, but one thing I'd like to learn to do is to be able to write in a less scholarly kind of way. So, I noted this South African scholar called Chabanyi Manganyi who starts off doing psychology. He's a clinical psychologist. And then somewhere in his career, he switches over and he starts writing biographies. I went to this biographer's conference in D.C. earlier this year and I heard Taylor Branch speak.

Branch writes a whole number of books on Martin Luther King and Civil Rights. And he gave a talk in which he stressed that there's some stories you can only tell in a narrative genre, there's some political stories that can only effectively be recounted in that way. So I found that inspiring. I don't know if I've got the talent or the ability to write like that. With my interest in Sobukwe, what I'd like to do is learn to be able to write biographically, because—following Branch—it seems that that's the most useful way to be able to tell stories like that of Sobukwe's political journey.

Actually, there's a little anecdote that springs to mind. A few months ago I read S. C. Gwynne's *Empire of the Summer Moon* (2010).

HM: Oh, I think I've heard of that work.

DH: It's a history of the Comanches. And one can imagine that the author had something of a political problem in thinking about how to tell this story. Why so? Well the Comanches are pretty war-like, and they did some brutal stuff. Of course, politically, it is problematic to retrieve that history and foreground in a certain way, because to do that is to facilitate a whole series of fantasies about what the American Indians actually were like and retroactively justify the genocidal violence done to them.

So how then do you then tell the story of the Comanches while simultaneously wanting to shy away from using reductive stereotypes and all the racialized fantasies that inevitably follow? It is a politically problematic situation, because you don't want simply to ignore parts of the recorded history, but it is bad politics to foreground that stuff because then you're on a one-way ticket to savagery, barbarism, etc.

HM: Or neocolonialism.

DH: So how do you do the story? The way Gwynne does it—and I think he does it successfully—is he doesn't leave the violence out, he includes it. But part of the book starts off as a history or—yeah, a history of the Comanche people, but then it zeroes in on this guy, Quanah, who's a chief right when the Comanche tribe is facing its most severe challenges of extinction.

He tells us so much about Quanah's story that we get to know him a little bit. And Quanah does some questionable stuff, he certainly scalps people. But through the course of the story, you see how Quanah is able to change, how he's able to adapt. And by the end of the book, you don't—well, it depends, presumably, a little bit on the reader—you don't easily adopt the view that Comanches are simply somehow barbaric, a violent people. And I think Gwynne can only really manage to do that by being skilled as a narrative author, and by being able to tell a personalized story.

And hence, that would be one answer to the question. Whether I'm able to do it, I don't know. But to learn a little bit about biographical writing. And interestingly, I think– I don't know how to thematize that or articulate it, but I think that is part of what an alternative psychology would be anyways, right?

HM: Yes.

DH: About a successful way of being able to manage and work with narratives and tell those kinds of stories. About doing justice to human experience.

HM: Well, in some ways what you're describing solves the dilemma you alluded to earlier, where we—there is only one race, and that is the human race. And yet, there are all these issues of injustice. And it seems the psycho biographical approach can unify those themes.

It is time for us to end but thank you so much for your thoughts during this interview.

Bibliography

Biko, S. (1978). *I write what i like*. London: Bowerdean.

Blackman, L., Cromby, J., Hook, D., Papadopoulos, D., & Walkderdine, V. (2008). Creating subjectivities. *Subjectivity*, *1*, 1–27.

Bowman, B., & Hook, D. (2011). Paedophile as apartheid event: Genealogical lessons for working with the Apartheid Archive. *Psychology in Society*, *40*, 64–82.

Butler, J. (1997). *The psychic life of power: Theories in subjection*. Stanford, CA: Stanford University Press.

Coetzee, J.M. (1999). *Disgrace*. London: Vintage.

Crapanzano, V. (1985). *Waiting: The Whites of South Africa*. London & New York: Granada.

Fanon, F. (1967). *Black skin, White masks*. New York, NY: Grove Press.

Fanon, F. (1986). *The wretched of the earth*. New York, NY: Grove Press. (Original work published 1963).

Foucault, M. (1977). *Discipline & punish the birth of the prison*. London: Allen Lane.

Gholizadeh, S., & Hook, D. (2012). The discursive construction of the 1978 Iranian Revolution in the speeches of the Ayatollah Khomeini. *Journal of Community and Applied Social Psychology, 22*(2), 174–186.

Gortimer, N. (1980). *Burger's Daughter*. New York: Penguin Books.

Gwynne, S.C. (2010). *Empire of the summer moon*. New York: Scribner

Hook, D. (2001a). Critical psychology in South African: Applications, limitations, possibilities. *Psychology in Society, 27*, 3–17.

Hook, D. (2001b). Discourse, knowledge, materiality, history. *Theory & Psychology, 11*(4), 521–547.

Hook, D. (2001c). The disorders of discourse. *Theoria, 48*, 41–68.

Hook, D. (2001d). Therapeutic discourse, co-construction, interpellation, role-induction; Psychotherapy as iatrogenic treatment modality?. *The International Journal of Psychotherapy, 6*(1), 47–66.

Hook, D. (2002). The other side of language: The body and the limits of signification. *Psychoanalytic Review, 89*(5), 681–713.

Hook, D. (2003a). Analogues of power: Reading psychotherapy through the sovereignty-discipline-government complex. *Theory & Psychology, 13*(5), 605–628.

Hook, D. (2003b). Foucault's psychoanalysis. *Psycho-analytic Psychotherapy, 11*(2), 24–39.

Hook, D. (2003c). Language and the flesh: Psychoanalysis and the limits of discourse. *Pretexts: Literary and Cultural Studies, 12*(1), 43–64.

Hook, D. (2003d). Psychotherapy and 'ethical sensibility.' *International Journal of Psychotherapy, 8*(3), 195–212.

Hook, D. (2003e). Reading Geldenhuys: Constructing and deconstructing the Norwood Killer. *South African Journal of Psychology, 33*(1), 1–9.

Hook, D. (Ed.). (2004a). *Critical psychology*. Cape Town, South Africa: University of Cape Town Press.

Hook, D. (2004b). On the psychical technology of monuments. *Free Associations, 11*, 400–421.
Hook, D. (2004c). The psychoanalytic pastoral. *Psycho-analytic psychotherapy, 12*(1), 19–36.
Hook, D. (2004d). Psychotherapy: Elements from a history of critique. *Journal of Critical Psychology, Counselling and Psychotherapy, 4*(3), 135–154.
Hook, D. (2004e). Racism as abjection: A psychoanalytic conceptualization for a post-apartheid South Africa. *South African Journal of Psychology, 34*(4), 672–703.
Hook, D. (2005a). Affecting whiteness: Racism as technology of affect. *International Journal of Critical Psychology, 16*, 74–99.
Hook, D. (2005b). A critical psychology of the postcolonial. *Theory and Psychology, 15*(4), 475–503.
Hook, D. (2005c). Genealogy, discourse, 'effective history.' *Qualitative Research in Psychology, 2*, 1–29.
Hook, D. (2005d). Monumental space and the uncanny. *Geoforum, 36*, 688–704.
Hook, D. (2005e). On subjectivity, the historicity of discourse and questions of power-knowledge and 'politics. *Qualitative Research in Psychology, 2*, 34–38.
Hook, D. (2005f). Paradoxes of the other. *Psychology in Society, 31*, 9–30.
Hook, D. (2005 g). The racial stereotype, colonial discourse, fetishism, racism. *The Psychoanalytic Review, 92*(5), 701–734.
Hook, D. (2006). 'Pre-discursive' racism. *Journal of Community and Applied Social Psychology, 16*, 207–232.
Hook, D. (2007). *Foucault, Psychology and the Analytics of Power*. London, UK: Palgrave.
Hook, D. (2008a). Absolute Other: Lacan's 'Other' as adjunct to critical social psychological analysis?. *Personality and Social Psychology Compass, 2*(1), 51–73.
Hook, D. (2008b). Articulating psychoanalysis and psychosocial studies: Limitations and possibilities. *Psychoanalysis, Culture & Society, 13*, 397–405.
Hook, D. (2008c). Fantasmatic transactions: On the persistence of apartheid ideology. *Subjectivity, 24*, 275–297.
Hook, D. (2008d). Postcolonial psychoanalysis. *Theory & Psychology, 18*(2), 269–283.
Hook, D. (2008e). The 'real' of racializing embodiment. *Journal of Community and Applied Social Psychology, 18*(2), 140–152.

Hook, D. (2009). Lacan's Kantian logic of sexuation. *Journal of the Centre of Freudian Analysis & Research, 19*, 89–118.
Hook, D. (2010a). Power and das Ding: Sublimation as a political factor. *Journal of the Centre of Freudian Analysis & Research, 20*, 91–121.
Hook, D. (2010b). The powers of emptiness. *Theory and Psychology, 20*(6), 855–870.
Hook, D. (2011a). *A critical psychology of the postcolonial: The mind of apartheid.* London, UK: Routledge.
Hook, D. (2011b). Narrative form, 'impossibility' and the retrieval of apartheid history. *Psychoanalysis, Culture & Society, 16*, 71–89.
Hook, D. (2011c). Psychoanalytic contributions to the political analysis of affect and identification. *Ethnicities, 11*(1), 107–115.
Hook, D. (2011d). Retrieving Biko: A Black Consciousness critique of whiteness. *African Identities, 9*(1), 19–32.
Hook, D. (2011e). White privilege, psychoanalytic ethics and the limitations of political silence. *South African Journal of Philosophy, 30*(3), 495–502.
Hook, D. (2012a). Apartheid's lost attachments (1): On psychoanalytic reading practice. *Psychology in Society, 43*, 40–53.
Hook, D. (2012b). Apartheid's lost attachments (2): Melancholic loss and symbolic identification. *Psychology in Society, 43*, 54–71.
Hook, D. (2012c). Screened history: Nostalgia as defensive formation. *Journal of Peace Psychology, 18*(3), 225–239.
Hook, D. (2013a). Discourse analysis: Terminable or interminable?. *Qualitative Research in Psychology, 10*(3), 249–253.
Hook, D. (2013b). Nixon's 'full speech': Imaginary and symbolic registers of communication. *Journal of Theoretical and Philosophical Psychology, 33*(1), 32–50.
Hook, D. (2013c). *(Post)apartheid conditions.* London, UK: Palgrave.
Hook, D. (2013d). The racist bodily imaginary: The image of the body-in-pieces in (post)apartheid culture. *Subjectivity, 6*, 254–271.
Hook, D. (2013e). Towards a Lacanian group psychology: The prisoner's dilemma and the trans-subjective. *Journal for the Theory of Social Behaviour, 43*(2), 115–132.
Hook, D. (2013f). Tracking the Lacanian unconscious in language. *Psychodynamic Practice, 19*(1), 38–54.
Hook, D. (2014a). Antagonism, social critique and the "violent reverie.". *Psychology in Society, 46*, 21–34.

Hook, D. (2014b). Love, artificiality and mass identification. *Psychodynamic Practice, 20*(2), 128–143.
Hook, D. (2014c). *Steve Biko: Voice of liberation*. Cape Town, South Africa: HSRC Press.
Hook, D. (2015). Petrified life. *Social Dynamics A Journal of African Studies, 41*(3), 438–460.
Hook, D. (2016a). Of symbolic mortification and "undead" life. *Psychoanalysis & History, 18*(2), 221–256.
Hook, D. (2016b). Sobukwe and the psychosocial. *Psychology in Society, 50*, 9–34.
Hook, D. (2016c). A threatening personification of freedom or: Sobukwe and repression. *Safundi: The Journal of South African and American Studies, 17*(2), 1–24.
Hook, D. (2016d) A threatening personification of freedom or: Sobukwe and repression. *Safundi, 17*(2), 189–212.
Hook, D., & Harris, B. (2000). Discourses of order and their disruption: The texts of the South African Truth and Reconciliation Commission. *South African Journal of Psychology, 30*(1), 14–22.
Hook, D., & Howarth, C. (2005). Future directions for a critical social psychology of racism/antiracism. *Journal of Community & Applied Social Psychology, 15*(6), 506–512.
Hook, D., & Long, C. (2011). The Apartheid Archive Project, heterogeneity, and the psychoanalysis of racism. *Psychoanalysis, Culture & Society, 16*, 1–10.
Hook, D., & Neill, C. (2008). Perspectives on Lacanian subjectivities. *Subjectivity, 24*, 247–255.
Hook, D., & Neill, C. (2010). Žižek, political philosophy and subjectivity. *Subjectivity, 3*(1), 1–6.
Hook, D., & Parker, I. (2002). Deconstruction, psychopathology and dialectics. *South African Journal of Psychology, 32*(2), 49–54.
Hook, D., & Truscott, R. (2013). Fanonian ambivalence: Psychoanalysis and postcolonial critique. *Journal of Theoretical and Philosophical Psychology, 33*(3), 155–169.
Hook, D., & Vanheule, S. (2015). Revisiting the master signifier, or, Mandela and repression. *Frontiers in Psychology: Psychoanalysis and Neuropsychoanalysis*. doi: 10.3389/fpsyg.2015.02028.
Hook, D., & Vrdoljak, M. (2002). Gated communities, heterotopia and a "rights" of privilege. *Geoforum, 33*, 195–219.

Hook, D., Franks, B., & Bauer, M. (Eds.). (2011). *Social psychology of communication*. London, UK: Palgrave.
Howarth, C., & Hook, D. (2005). Towards a critical social psychology of racism: Points of disruption. *Journal of Community & Applied Social Psychology, 15*(6), 425–431.
Mbembe, A. (2001). *On the postcolony*. Berkeley, CA: University of California Press.
Parker, I., & Hook, D. (2008). Psychoanalysis and social psychology: Historical connections and contemporary applications. *Journal of Community and Applied Social Psychology, 18*(2), 91–95.
Pietrusza, C., & Hook, D. (2016). "You're kicking me out?" Scansion and the variable-length session in Lacanian clinical praxis. *Psychodynamic Practice, 22*(2), 102–119.
Polatinsky, S., & Hook, D. (2008). On the ghostly father: Lacan on Hamlet. *Psychoanalytic Review, 95*(3), 359–385.
Ratele, K., & Hook, D. (Eds.). (2004). *Self, Community & Psychology*. Cape Town, South Africa: University of Cape Town Press.
Rienstra, B., & Hook, D. (2006). Weakening Habermas: The undoing of communicative rationality. *Politikon, 33*(3), 313–340.
Stevens, G., Duncan, N., & Hook, D. (Ed.) (2013). *Race, memory and the Apartheid Archive*. London, UK: Palgrave.
Terre Blanche, M., Bhavnani, K., & Hook, D. (Eds). (1999). *Body politics: Power, Knowledge and the body in the social sciences*. Johannesburg, South Africa: Histories of the Present Press.
Žižek, S. (1997). *The plague of fantasies*. London and New York: Routledge.

Heather Macdonald is Assistant Professor of Psychology at Lesley University. She came to academia after years of practice as a clinical psychologist whose work involved community outreach, child assessment, and individual therapeutic services to children and families in the foster care system and with youth involved in the juvenile justice system. Her work in urban environments and abroad has led to scholarly research on the interface between culture, social justice, relational ethics, clinical practice, and postcolonial thought. Her first monograph, titled *Cultural and Critical Explorations in Community Psychology* (2017) published by Palgrave Macmillan, further considers the implications of psychological assessment, diagnosis, and historical trauma. As much as the book represents a cultural critique of more traditional clinical discourse, it also suggests that if we view clinical relationships from the perspective of a relational ethics, that includes discourse about cultural memory, freedom, power, and other events that break with the neoliberal expectation of "mutual reciprocity," then therapeutic work can also be a site of transformation and reversal that moves beyond fixed racial categories. She is a fellow of the Psychology and the Other Institute.

Psychoanalytic Sensibility and Honoring Individual Differences: A Conversation with Nancy McWilliams
(Interviewed by Marie Hoffman)

Nancy McWilliams teaches at Rutgers University's Graduate School of Applied and Professional Psychology, and has a private practice in Lambertville, New Jersey. She is author of: *Psychoanalytic Diagnosis*, (1994, rev. ed. 2011), *Psychoanalytic Case Formulation* (1999), and *Psychoanalytic Psychotherapy* (2004), all with Guilford Press. She has edited or contributed to several other books, and is associate editor of the *Psychodynamic Diagnostic Manual* (PDM, 2006; PDM-2, 2017).

Dr. McWilliams is former president of the Division of Psychoanalysis, Division 39, of the American Psychological Association, and is on the editorial board of *Psychoanalytic Psychology*. A graduate of the National Psychological Association for Psychoanalysis, Dr. McWilliams is also affiliated with the Center for Psychotherapy and Psychoanalysis of New

N. McWilliams (✉)
Rutgers University, New Brunswick, NJ, USA
e-mail: nancymcw@aol.com

H. Macdonald et al. (eds.), *Dialogues at the Edge of American Psychological Discourse*, Palgrave Studies in the Theory and History of Psychology, DOI 10.1057/978-1-137-59096-1_11

Jersey and the National Training Program of the National Institute for the Psychotherapies in New York City.

Dr. McWilliams won the Gradiva prize in 1999 and the Goethe Scholarship Award in 2012; the Rosalee Weiss Award for contributions to practice in 2004; the Laughlin Distinguished Teacher Award, 2007; the Hans Strupp Award for teaching, practice and writing in 2014; and the Division 39 Awards for both Leadership and Scholarship. She was the Erikson Scholar at the Austen Riggs Center in Stockbridge, MA, in the summer of 2016.

She has given commencement addresses for the Yale University School of Medicine and the Smith College School for Social Work. The American Psychological Association chose her to represent psychoanalytic therapy in a 2011 remake of the classic film, *Three Approaches to Psychotherapy*, and asked her to be a plenary speaker for the 2015 APA convention in Toronto.

She is an honorary member of the American Psychoanalytic Association, the Moscow Psychoanalytic Society, and the Warsaw Scientific Association for Psychodynamic Psychotherapy. Her writings have been translated into 20 languages.

Marie Hoffman (MH): Could you tell us a little bit about your personal history and about what ultimately drew you specifically to psychoanalysis as your preferred orientation in this overall field of psychology?

Nancy McWilliams (NM): Well, I think probably everything conspired to make me into a psychologist. My mother was very psychological. She died when I was nine, so I don't know too much about how she viewed her own psychological mindedness, but she was a graduate of the Columbia Teachers College master's program in education, and she taught the deaf. I've been told that the programs at Teachers College were highly influenced at that time, the 1930s, by psychoanalysis. I remember many little lessons in empathy and psychology and how to understand other people from her.

My father was a loving father and a good father in most ways, but difficult. He would have rage outbursts. As a result, he kept me fascinated [laughter] and a little scared. I didn't figure out until I was an adult that what I was seeing was the effect of brain damage as a result of his having had encephalitis lethargica. He had the sleeping sickness, which Oliver Sacks (1990) wrote about in *Awakenings*. So my interest in diagnosis, I think, came from my trying to figure out my father.

Then in my teen years, after my mother's death, I was very influenced by a lesbian couple in the 1950s and 1960s, who ran a Girl Scout camp that I went back to every summer when my family was moving around a lot. That camp gave me a lot of stability and continuity. They were very psychological and had been influenced, in the case of one of the two partners, by Theodore Reik, whose writings were popular at the time.

Then at Oberlin College, I majored in political science, because in psychology you had to run rats for two years before you could talk about people. The political scientists were talking about people from the get-go.

In my junior year, my faculty advisor, whom I later married, gave me a book by Freud (1930/2010), *Civilization and Its Discontents*, and suggested that because I was very psychologically minded, I might want to do my political theory honors thesis on Freud's political thought. That was my introduction to Freud, and I found him absolutely fascinating. I started reading psychoanalytic stuff, including Theodore Reik's books that he had around his house.

I married him my senior year, and we went to New York the following year. I started taking psychology courses because I decided I wanted to be a therapist. It's very temperamentally comfortable for me. I was always interested in people and individuality.

I realized Theodore Reik was still alive and that there was somebody I could talk to who had known Freud. I wrote to him and asked him if he would meet with me for an hour. I asked his advice on what a beginning psychologist should pay attention to, and he told me that to be a good therapist, I had to go through analysis. He directed me to his institute, where I was lucky enough to—lucky and rich enough—I could afford $15 an hour, which was, in 1969, the lowest price at which you could get an experienced analyst— to get a fine, senior analyst. They assigned me to a guy. I took potluck. He turned out to be a terrifically good match.

Although I approached my analysis originally as just a professional opportunity to understand more— in other words, I took a very intellectual attitude toward it—it greatly transformed my life. I am quite sure my marriage would not have lasted without it, because faculty/student marriages start off with idealizations that have to be worked through. I got a lot of help with that.

I'm quite sure I would not have had children, because at the time, it was a little bit before the feminist resurgence and I thought if I was an ambitious woman, which I was, I probably couldn't do everything. But in analysis, I discovered that I had a very powerful, unconscious belief that if you become a mother you die. My father had remarried after my first mother's death, and my stepmother, with whom I had a very good relationship, died while I was in college. So I watched two mothers die of cancer, and that made a deep impression on my amygdala, evidently. If I hadn't made that unconscious conviction conscious, I don't think I would have had kids. Which would have been, from my perspective now as a mother of two lovely women and grandmother of three, a tragedy.

So, it was really the unexpected transformative experience of my own psychoanalysis that led me to be psychoanalytically oriented. I originally went into it with the idea, "That's what the old guys used to do; let's learn about that and then we'll move on to what's more helpful."

But it was extremely helpful. At Rutgers, where we went because of my husband's job offer, I didn't specialize in clinical psychology because there were very few psychoanalytically friendly people there. But in personality psychology, there was Silvan Tomkins. Then the second year I came, George Atwood was hired. The third year I was there, Bob Stolorow joined the faculty. They began their conversations. So I was in on the ground floor of the intersubjective movement there. My department chair kind of liked the Kleinians and the object relational literature. So I found a psychoanalytically very friendly space at Rutgers.

Everything sort of conspired, as I said, to make me not just a psychologist but a psychoanalytic psychologist.

MH: What fascinating history. Thank you for sharing that. I imagine everything conspiring in terms of your history and your training also caused you to be impassioned about certain areas. So I'm wondering if you could summarize what those areas are that impassioned you most as you've gone into research and work as a result of all that you've shared.

NM: Well, first let me correct you on something. I'm not a researcher. The only research I ever did was for my master's thesis, and although it was interesting, and although I was trained to do research, and I did chair a number of research doctoral theses later, I am a therapist, I'm a teacher. But not a researcher.

But your question was, what are my passionate concerns?

MH: Yes, the areas that you really feel passionate about.

NM: Appreciating individual differences and not presuming that you know other people's psychology before you let them teach you about it. I've gotten a reputation because of my textbook on diagnosis for being invested in a broader kind of diagnosis than the DSM. But nevertheless, I'm associated with diagnosis. I find that a little bit odd because although it's understandable, I'm as interested in the implications of someone's ethnicity, religious background, positioning in the family, socioeconomic level, whether they're a twin, whether they're an adoptee, whether they have some kind of physical disability, and what are the implications of that—of race, of being in a big family versus a small family.

I think to understand people as well as possible, which is never completely, there are so many different factors you have to take into account—their temperament, their attachment style, their defensive style, their affect, their unconscious beliefs. I'm really just interested in all the unique ways that all of us, who have so much in common and share the planet, differ in subtle ways from each other and misunderstand each other.

So that's my passion.

MH: We were just talking before the interview about your worldwide travels, and I'm wondering if you have enriched that passion of yours to detect and understand individual differences as you've gone to various parts of the world. Perhaps you can give a couple of examples.

NM: Yes. It's wonderful to learn from therapists about what they see as some of the psychological issues for people in their countries and subcultures. This is true in the United States and North America, as well as in other countries. I come from a long line of teachers. From my first course in 1970, I've been trying to find ways of passing on interesting things that I've learned, especially things that make it easier to understand other people. So my audience, as I think of it, has never been the psychoanalytic community. I've been taking ideas that came out of the psychoanalytic community that are useful and passing them on to other people.

So my typical audience, when I go and speak anywhere, including in the United States, is not likely to be psychoanalytic institutes, although occasionally I talk at those. It's more likely to be a very broad range of people in all kinds of human service roles, not just doing private practice, but maybe working at counseling centers, in jails, at drug treatment centers, and doing various charitable kinds of work.

So I speak to a broad range of therapists in, as it's happened, a broad range of countries. Although they all recognize the categories in, let's say, my diagnosis book, as describing some people that they know, they'll tell me that the frequency or the loading is different, depending upon their culture.

I've taught in some collectivist cultures that are quite different from ours, including China, Iran, Turkey, Singapore, and South Africa, including some tribal groups. We have a lot more in common than we have differences, but they have educated me about some different sensibilities, too.

You asked for an example. One would be from when I was teaching in China a year ago. I got a question from the Chinese therapists that you would never get from an American audience. They wanted to know how to help their depressed grannies. The Chinese female therapists explained to me that their grandmothers had been brought up in a culture in which they'd been trained to first obey their father, then obey their husband, then obey their son. Now they're watching their granddaughters with sexual freedom, with job autonomy, with a sense of agency, and they're envious and get depressed.

I was struck, thinking I can't imagine American audiences as a group asking me how to help their grandmothers. But in China, of course, with the sense of responsibility for the previous generations, that's a question in the front of their hearts.

So that was fascinating.

MH: That is a fascinating example. You mentioned that by and large you speak to a much more general audience of people helpers than simply the psychoanalytic community. So, I'm wondering how the field of psychology has responded to your work.

NM: I would say mostly in a quite friendly way. I think that psychoanalysis in general is very devalued by most contemporary psychologists. But I haven't felt personally devalued. And for people of other orientations who read my work, I think they have found it useful. Marsha Linehan was very friendly to me and even told me that she thought I was an exceptionally good therapist.

So on the personal level I haven't felt treated prejudicially. Quite the contrary, like the APA publishing group that asked me to do the reprise of the "Gloria" film, (Rogers & Shostrom, 1965), they seemed quite happy to have a representative of psychodynamic therapy as one of the three main evidence-based orientations, and that's how they framed it.

So this has been true for me even though there are plenty of people in APA who are anti-psychoanalytic. Worse than that, I would say there's a general set of presuppositions in APA that are anti-humanistic, in some ways anti-scientific, because their definition of science is so incredibly narrow that it's a threat to the kind of work that I think is most valuable.

In general, I've had only positive experiences with groups that have different orientations from mine.

MH: As I've experienced you in different settings, I've had the privilege of noting that as well. Your work is very positively received. So if we were to look at any resistances or impediments to your work in the contemporary discipline of psychology, are there resistances specific to your being a psychoanalyst in a general psychological audience?

NM: Well, there are certainly some researchers who don't have much clinical experience, who define psychoanalysis as a technique. I don't define it as a technique. I define it as a body of knowledge and a whole group of techniques that flow from a certain theoretical sensibility. But if you define it as a technique and you define that technique as putting people on the couch, telling them to free associate and following a very rigid pattern with them, it's easy to devalue it. There are many researchers who think that that is psychoanalysis, or who think that what Freud said in 1913 defines contemporary psychoanalysis and psychoanalytic psychology.

That is a kind of misunderstanding that pervades APA, and I feel frustrated that there aren't more people speaking up against it. Psychoanalytic people, I guess because we get exhausted with trying to defend ourselves, like to talk to each other. We haven't talked enough to those of our colleagues who don't have as much, if any, clinical experience. And their relative lack of experience is the result of what's happened in academia. You'd be crazy to have a private practice now if you want to be a researcher and get tenure. You just don't have the time. It's not that they wouldn't want to do that; it's a death sentence to your career if you want to be a certain kind of academic.

So they don't have empathy with what it's like to be a therapist. I would also say that therapists don't really have an empathic attitude about what it's like to be a researcher either, or what are the stresses on academics. We both tend to "other" each other. I would say currently that the rift between academic researchers and practitioners, as a group, is a more problematic rift than the rift among people of different orientations within psychology.

MH: That's a very interesting observation and leads us right to a broader question. If you had to speak prophetically to the field of psychology, what would you feel it would need to do to right its course, to be more balanced, to accomplish its mission?

NM: Well, it depends on how you define its mission. The American Psychological Association has in general defined its mission as increasing the power of psychologists in various existing powerful bodies in the culture. I think that has been a shame, and it's led us down the path of colluding with torture, in only the most dramatic recent example. But it's been throughout the field of institutional psychology.

We supported in most respects the whole eugenics movement. In fact, the Nazis got some of their arguments from arguments that were made by American psychologists. The architect of the apartheid system in South Africa, Hendrik Verwoerd, was a psychologist who used his knowledge, much of which came from American social psychology, to persuade a large public about white supremacy.

The behaviorists had a very negative effect on child rearing in the United States that lasted for at least two generations. I mean the Watsonian behaviorists, not the contemporary cognitive behavior therapists.

The IQ movement had various kinds of casualties, including marginalizing people from subcultures that didn't share quite the same knowledge base with the dominant cultures in the United States.

I think psychologists in general are much more interested, at least I would hope, in finding what is true rather than in having increasing clout in existing organizations. That's where we need to rethink ourselves organizationally. I think this torture scandal gives us an opportunity to do that. I have to say I don't have much optimism about it because it's very hard to change institutional culture. Even though people in power in APA are very distressed about what they've learned, and very concerned to right the ship, there are so many ways in which all the reinforcements are on the side of the narcissistic pursuit of organizational importance, rather than the sheer pursuit of knowledge.

Psychologists are not alone in this problem. A similar process is happening everywhere. I was just hearing a program on cancer research that was talking about how in the '70s, research was done for its own sake, not with the immediate concern for what was the application going to be. Now with the influence of the drug companies and insurance companies, we find that most of the grants that go to universities for research are implicated in drug company agendas. So, we've got a contamination of the traditional mission of the university with a lot of corporate agendas. It seems to me it will be very hard to extract ourselves from that.

MH: You seem to be saying that the first step in responding prophetically would be to define what the mission of psychology is. So if you could put it in positive terms, how do you see the mission of psychology?

NM: Well, first of all understanding should precede application. We move very fast to application. Of course, that is a psychoanalytic perspective. I don't think of people in terms of social phobia or eating disorders. I think of them in terms of who is this person who has suffered this affliction? I think we've been very quick to jump instead to, "How do you describe this category of affliction and what specific approach can we aim at it?" I would say that conflates a certain type of research with what clinicians have traditionally done, which is to try to understand people in as much depth as possible, and then your technique flows from that.

The reason I say it's conflated with research is that if you want to do a certain kind of outcome research, you have to define what you're interested in very clearly. You have to take measures on it. You have to manualize what you do so that all participants are doing the same thing that is being assessed. You have to do a short-term intervention on the question, because your grant doesn't last forever. You have to judge improvement based on symptom reduction, because that's what is measurable in a short period of time. Then you go from there.

In therapy, we take whoever comes in the door. They usually don't have one problem not comorbid with anything else; that's a kind of cherry-picking that researchers have to do in order to be clear what they're looking at. We tend to be able to allow some time to pass so that we get a deeper understanding of the person. We don't need to manualize; that's a research requirement, it's not a clinical requirement.

We judge improvement not just by symptomatic improvement, but by larger frameworks such as "Is the person's life improving in the areas of love, work, and play?" Or increasing in self-esteem, and affect tolerance and regulation, or security of attachment? Is the person developing the capacity to accept what can't be changed and grieve and move on? Does the person have a sense of vitality? All of those things are more important from the perspective of therapists than tinkering with a discrete symptom.

So I'd like to pull us back to that way of thinking. My next project is an effort to do that. It's a book on that kind of overview of what we are trying to work *toward*. What is the good life? What is psychological wellness? What is mental health?

MH: You address the myopia that can result by simply looking at symptoms and applications. Yet I've also read, in psychoanalytic writing that you've done, that you promote some degree of research in psychoanalysis. You don't seem to do a total reversal, that is, that we can't have any research. There seems to be a real balance in what I've read, and I'd like you to speak to that.

NM: Well, I think research has gotten defined very narrowly as doing research on specific techniques. Although I think we've learned some valuable things from that kind of research—exposure therapies, for example, for certain kinds of suffering—the kind of research that most interests me is research on personality differences, on attachment patterns, on affective transmission, on neuroscience, on the implications of divorce on children, for example—mostly research that requires long-term work that is hard to support on grants.

I'm very interested in that long-term Harvard study of those men who first were measured in the '50s; now some of them are in their 90s, I believe. I think longitudinal research is terribly valuable, not just these little snippets of what's easy to research. Unfortunately, again, the reinforcements in the field are all for getting a whole list of small pieces of research out there.

Some of the research that people like Peter Fonagy's research group and Otto Kernberg's research group are doing, on approaches to borderline psychology, has been very valuable. By its very nature, that subject matter does not admit of short-term work. I mean, you can change some symptoms in the short term, but not really change the personality.

I'd like to see a lot more research like that. I'm fascinated by research that illuminates something important about human psychology. I think we have to support researchers. It was psychoanalysts' indifference to research that gave us a terrible name among mainstream psychologists, because we were arrogantly saying, "Well, if you'd been *analyzed*, you would know this; you wouldn't have to do research on it." And that's an anti-scientific, contemptuous attitude that characterizes, in my experience, only a small number of psychoanalysts. But there were enough of them that we got a terrible reputation on the basis of such comments.

MH: I really appreciate the balance that you bring to that whole topic. Once again, what you first clarify is the definition, whether it's definition of mission or the definition of research. It's so important to understand how we're using the word. You already spoke to this somewhat, and it's very interesting, as we're talking, that you're picking up on a thread that is going to be emerging in one of the questions. So I will pose the question and you can add perhaps even more than what you've already said.

What are the historical, political and/or theoretical concerns in psychology today that you feel have the most influence on the future of contemporary psychological practice?

NM: Well, historically—let me take them one at a time. Historically, I think I can mostly speak for American psychology. I think we've become narrower in many ways over the past decades. Partly because of the dying off of all the European immigrants who enriched our way of thinking. America is in some ways a somewhat adolescent culture. We want to fix things. We want to believe we have infinite resources. We want to just be pragmatic and practical and get it done.

Historically, as the Holocaust era immigrants—in all kinds of fields, not just psychology—are dying off and losing their influence, we're seeing a resurgence of a sort of simple-minded American practicality, which in some ways is lovely; it's a nice part of our culture, the problem-solving part of our culture. But it's exploited by companies that don't want to pay for psychotherapy, who then say, "Just fix the symptom and get the guy back to work," rather than, "Let's really improve this guy's life so he won't ever suffer this way again."

So historically, I think we're suffering what all huge democracies suffer, which is increasing bureaucratization and the sense that our record-keeping becomes more important than our actual work, for example. It's not different from what teachers are complaining about in teaching to the test, rather than really inspiring kids to be more curious, more able to critique an argument, more able to make an argument. It's not different from what doctors are experiencing when they are pressed to do worthless tests in order to keep themselves litigation-free, or when they're pressed *not* to do tests because of expense, because of the insurance company's short-term concerns. They're having to fight with bureaucrats to take care of their patients, too.

This is happening in all kinds of areas. It's really an artifact of bigness. We're not a small community where we know each other and can trust each other in many respects. So we have to have all these mechanisms for going through various hoops. I think it's deadening.

Politically, we've certainly seen global capitalism remove, from many people, the sense that they have agency over their lives, and that has many psychological effects. Many people find it hard to find work that they think has meaning, for example. They can find work, but it doesn't necessarily have meaning. It's much harder in a mass culture, and now in a global culture, to feel like one matters at all, that one's efforts are going to be fruitful in some important way.

So politically, I think we're all feeling very small and helpless.

Theoretically, the cognitive-behavioral movement has had a huge influence on standard practice. Some positive, some negative, from my perspective. I have to say that my cognitive behavioral colleagues who are more or less full-time and experienced therapists have the same reservations I do about the pressures we're all under to do things very fast, according to pre-formulated rituals.

I think the relational movement has been very valuable to the psychoanalytic tradition in many ways. It's an egalitarian kind of movement, away from an authoritarianism that crept into psychoanalysis.

But you're asking about contemporary psychological practice in general. I would say that the kind of work that you and I do is getting harder and harder to do. My students have a much harder time starting a practice than I did almost 40 years ago. Our national organization has not supported private practice. It's been much too interested in getting on the bandwagon on integrated practice and some other agendas, like prescription privileges and things of that nature. Which may be valuable, but I think is taking resources away from supporting those people who are in the trenches, trying to help people, and trying to have some autonomy over their own lives.

I think psychoanalytic practice will survive, but I worry very much that it will survive mostly outside the healthcare system and mostly with people who can afford psychotherapy, not with working-class, middle-class and poorer people who used to be able to be helped by therapists back in the glory days when we had insurance policies that covered what I think of as real psychotherapy, of analytic and other types.

So I'm a temperamental optimist, but I'm actually an intellectual pessimist about where we're going in the field.

MH: It's refreshing to hear how you put the human person at the center of everything that you think about and do. It's really beautiful. The human person lives in a context, and you've touched on some of what I'm going to ask, but I'd like you to elaborate.

Historically, the practice of psychology has not associated itself with any specific forms of political activism or social responsibility. Has psychoanalysis been any different? And in your own work, how have you responded to that issue?

NM: Well, I think I would reject the premise of that question, first of all. Because I think the practice of psychology has been very affiliated with certain political agendas. They're just less visible. But if you try to take those on—for example, we've trained generations of people in how we do IQ testing. That had all kinds of social effects. It was a political position to take that position. It was a political position to get on board with the managed care movement instead of fighting it years ago when it started up. It was a political position to want to be important to the military.

So I think we have been associated with political activism. It just hasn't been activism against the powers that be. You can't live your life without making political choices all the time. Whether or not you're aware that that's what you're doing.

I think psychoanalysis, by virtue of being a movement that was not born in the United States, has had a kind of privileged position to look at the culture from outside in certain ways, much as de Tocqueville did, or even Herodotus, if you look at political theorists who can comment about cultures because they're not fully in them.

Psychoanalysts tend to be critical of some aspects of the culture that mainstream Americans take for granted. You certainly see that in the fact that it was largely Division 39 members in APA who kept trying to hold APA's feet to the fire about the issue of torture and collusion with military agendas, and what was happening to the ethics code.

So yeah, I think psychoanalysis has been somewhat different. I think, though, that there are probably prejudices within psychoanalysis that also have political implications. That was true, for example, when mainstream psychoanalysis thought that—and by the way, this wasn't really Freud's position—when they thought that homosexuality was pathological by its nature. That belief had enormous negative political effects on gay and lesbian people.

Every decision we make with the DSM (APA, 2013) has some political consequences. Whether you call somebody "on the autistic spectrum" or "Asperger's" has some consequences for that person. Whether we define certain kinds of things as insanity has an impact on the legal system.

Psychoanalysts in general have been a little bit outside the dominant culture, and in some ways that's healthy. We question. We ask whether adapting to this particular work role is really a satisfying way to live your life, rather than immediately assuming that you get the person back functioning on a job that's not satisfying, for example.

But I wouldn't say that we have been morally superior in a general way to other people who are concerned about politics and where the world is going.

MH: You are one of the principal movers behind the *Psychoanalytic Diagnostic Manual* (PDM Task Force, 2006; Lingiardi & McWilliams, 2017). That was very much one of your projects for many years. Following up on what you just said: to what extent do you feel that the PDM, in contrast to the DSM, may do a somewhat better job as far as not pigeonholing people, but treating them in a more human way?

NM: Well, we do need some kind of language to talk about individual differences. As soon as we get it, we run the risk of objectifying, essentializing, and othering people. That's the problem with any diagnostic concern.

I think the PDM does that less than the DSM. That was the effort with the PDM, to bring back a biopsychosocial kind of diagnostic sensibility, away from just simply neo-Kraepelinian categorical diagnosis that's useful for researchers, but not so useful for therapists.

Having said that, I see that it could be misused in the same way. You can diagnose somebody who's borderline with the PDM as well as you can diagnose them that way with the DSM, and it still has a negative consequence in many situations.

But diagnosis just means really trying to understand. It comes from the Greek. You understand thoroughly, or through something. I think we all think that way, whether or not we name it. Even people who critique diagnosis, per se, will find themselves thinking, "Well, that's pretty hysterical," or, "That's pretty paranoid." Because they were trained in some way of dividing up the human pie.

Some diagnoses, I don't see how you can possibly be entirely against diagnosis, because some diagnoses are hugely important. For example, if somebody's rage reactions are a result of a temporal lobe epileptic problem as opposed to a borderline personality structure, as opposed to a bipolar illness where their mania is manifested as rage, those are important distinctions to make.

So yeah, with the PDM, we tried to put personality more at the front. We tried to talk about levels of functioning where somebody could be, let's say, high on self-esteem, but maybe low on affect regulation, so that people aren't so unidimensionally conceived. We tried to talk about the subjective experience of symptoms.

For example, with a categorical diagnosis that is useful for researchers, we define anxiety as sweaty palms, rapid heartbeat, feeling of apprehension and various other things that you can observe. But we don't define it in ways that are useful to therapists to figure out: Is this annihilation anxiety? Is it separation anxiety? Is it moral anxiety? Is it signal anxiety? Is it post-traumatic anxiety? Those would have different treatment implications.

So we tried to get the subjective experience in there. What does it feel like to the patient? Does the patient feel all-alone? Does the patient feel they're fragmenting into a million pieces? Or does the patient feel worried that they're going to violate their moral code? Those are not insignificant clinically, even though the manifestations of anxiety externally are the same.

So that conversation was absent from the ICD (WHO, 2004) and the DSM. We wanted to bring it back.

MH: The nuance and the complexity of the human person is very important to you, and not a simple category.

NM: Yes.

MH: I want to come back to specifically your personal work and your thinking. As you look back, what have been the crucial changes to your way of thinking or your approach to the discipline of psychology over time?

NM: I hope I've gotten more humble over time. I started out, and I think this is kind of a normal way to start out, so I'm not too embarrassed by this, but I started out feeling like, "Psychoanalysis is what helped me; that's the right road to go down." I had to learn other angles of vision. And the more I learned, the better therapist I was.

There's a wonderful article by Roy Schafer (1979) that talks about that process of starting out within psychoanalysis, being of one particular sub-orientation—being a Kleinian or being a Sullivanian, whatever. However, as you grow professionally, you end up expanding and taking elements from other approaches. Maybe you find new names and subsume them into your approach, or maybe you just do a more eclectic synthesis of your own.

But I've been very influenced by the work of Carl Rogers and Rogerian and humanistic psychology, by behaviorism, by the cognitive psychology movement, by neuroscience and biological psychiatry. The more I learn, the more my own angle of vision is very insufficient.

So I hope I've gotten more humble and more respectful, and more able to learn from people who don't necessarily accept my premises, but might have something to teach me.

MH: With what you've learned and where you're at, what do you imagine are your future directions? You alluded to this, but can you elaborate more?

NM: Well, I'm 70 years old, and I'm very aware that I don't have a sense of unlimited time anymore. So I've been asking myself, what impact would I like to have in the next decade? What I think I would like to do is start a public conversation about what we're trying to help people *toward* with psychotherapy. What is the implicit image of the healthy person?

I think there's a hunger for this in the culture. We've seen that in the positive psychology movement, in the interest in mindfulness, in the interest in some Eastern religions. There's an effort to expand our conversation about psychology into what are we trying to go toward, rather than what symptoms are we trying to eradicate.

Clinical psychologists and other therapists have more than 100 years of conversation about this. What I am working on now is a book that tries to elaborate on the good life—what do we know about that? What not only makes people happy—that's been the major focus of the positive psychology movement—but what besides happiness is valuable for human beings? I think most therapists would say that the capacity to tolerate unhappiness, to feel emotional pain when it's realistic to do so, is as important as feeling happy. We would put an emphasis on tolerating the whole range of affects and not using defenses that dissociate some experiences from us.

There are ways in which we haven't had a conversation, I think, as a culture since the early 1970s about what are we trying to help people toward. Back in the '70s, there was a lot of interest in this. Martin Luther King (1967), when he spoke at the American Psychological Association, questioned the whole paradigm of adjustment to the dominant culture, and said he hoped that psychology would be bigger than that.

We had *One Flew Over the Cuckoo's Nest* (Kesey, 1963), which raised the question of who's crazier, the inmates or Nurse Ratched. We had Maslow's (1987, 1999) hierarchy of needs. We had Jahoda's (Jahoda, 1958) work on mental health. We had a number of people who were trying to define what is psychological wellness, or are we just trying to adapt people to a sick society? That was the formulation then.

We haven't had that conversation for a long time. And it's time that this generation thought about that. What kind of life is a life that's worth living?

MH: It is interesting that in terms of your future direction of thought, it is very much political-social activism. It is saying, wake up, in a certain way, and really think. Don't be part of the machine.

NM: Yes.

MH: Are there any other thoughts that you wanted to elaborate on in terms of the questions before we come to our final question?

NM: I don't think so. I do want to say how grateful I've been to have been able to have a profession where I've made a good living doing what I love to do, whether it's helping people with emotional pain, or teaching, or writing, all three of which I really enjoy.

I worry that it will be harder to put all that together in the future for people who are seekers like me, who have about an equal balance between wanting to do good and wanting to understand. Those two motives, for me, come together so beautifully in psychoanalytic practice. We tend to assume that feeling better correlates with understanding; for me, that's a perfect mixture of humane moral impetuses.

I worry that therapists are going to be more and more treated as, and expected to be, technicians rather than healers.

MH: You are thinking about the next generation of healers. So if you were to give one either cautionary or inspirational piece of advice to an aspiring healer, aspiring psychologist, what might that be?

NM: I think therapists, with very few exceptions, have very good hearts. If you have a good heart, you already have what is the most important therapeutic asset you'll ever have when you sit with somebody and try to help them. I would say listen to your own heart. Don't do something with a patient that some expert tells you is right, if it feels wrong. Go consult. Go find out why it feels wrong. Trust your own natural right-brain-to-right-brain capacity to love and build on that.

Therapists are a pretty self-critical bunch, and they're very vulnerable to thinking that they must be wrong. But I think the heart has its wisdom. In addition to learning as much as you can about various techniques, various theories, various knowledge bases, you can't become estranged from your own natural instincts.

MH: That's a wonderful piece of advice. I thank you for sharing that. I guess in conclusion, I would simply like to have this for the record with you, Nancy. When people ask me about Dr. McWilliams, I can honestly say to them that I have rarely met a clinician who, both in practice, standing up and speaking, and at a personal level, is authentic. You're one and the same. That kind of integrity is rare and is, in my opinion, what makes you an incredibly—well, a unique person. I thank you for the opportunity to have done this interview with you.

NM: Well, thank you for saying that. It's been a pleasure.

Bibliography

American Psychiatric Association. (2013). *Diagnostic and statistical manual of mental disorders: DSM-5*. Washington, D.C: American Psychiatric Association.

Atwood, G. E., McWilliams, N., & Carroll, S. (1983). Inventors of new selves. *American Journal of Psychoanalysis*, *43*, 245–259.

Fors, M., & McWilliams, N. (2016). Collaborative reading of medical records in psychotherapy: A feminist psychoanalytic proposal about narrative and empowerment. *Psychoanalytic Psychology*, *33*(1), 35–37.

Freud, S. (1930/2010). *Civilization and its discontents* (J. Strachey, Ed. & Trans.). New York: Norton, W. W. & Company, Inc.

Gazzillo, F., Lingiardi, V., Del Corno, F., Genova, F., Bornstein, R. F., Gordon, R. M., & McWilliams, N. (2015, April 13). Clinicians' emotional responses and *Psychodynamic Diagnostic Manual* adult personality disorders: A clinically relevant empirical investigation. *Psychotherapy*. Advance online publication. doi.org/10.1037/a0038799

Gordon, R. M., Gazzillo, F., Blake, A., Bornstein, R. F., Etzi, J., Lingiardi, V., McWilliams, N., & Tasso, A. F. (2015). The relationship between theoretical orientation and countertransference expectations: Implications for ethical dilemmas and risk management. *Clinical Psychology and Psychotherapy*, Wiley Online Library. doi: 10.1002/cpp.1951

Greenberg, L., McWilliams, N., & Wenzel, A. (2013). *Exploring three approaches to psychotherapy.* Washington, DC: American Psychological Association.

Huprich, S., McWillliams, N., Lingiardi, V., Bornstein, R. F., Gazzillo, F., & Gordon, R. M. (2015). The *Psychodynamic Diagnostic Manual* (*PDM*) and the *PDM-2*: Opportunities to significantly affect the profession. *Psychoanalytic Inquiry, 35*, 1–13.

Jahoda, M. (1958). *Current concepts of positive mental health.* New York: Basic Books.

Kesey, K. (1963). *One flew over the cuckoo's nest.* London, England: Penguin Publishing Group.

King, M.L. (1967). The role of the behavioral scientist in the civil rights movement. *American Psychological Association Monitor*, 30(1).

Kottler, J. A., & Carlson, J. (2009). *Creative breakthroughs in therapy: Tales of transformation and astonishment* (pp. 231–236). New York, NY: Wiley.

Kottler, J. A., & Carlson, J. (2014). *On being a master therapist: Practicing what you preach* ((various pages)). Hoboken, NJ: Wiley.

Lingiardi, V., & McWilliams, N. (Eds.) (2017). *Psychodynamic Diagnostic Manual, 2nd Ed. (PDM-2).* New York: Guilford Press.

Lingiardi, V., McWilliams, N., Bornstein, R. F., Gazzillo, F., & Gordon, R. M. (2015). *The Psychoanalytic Diagnostic Manual Version 2* (*PDM-2*): Assessing patients for improved clinical practice and research. *Psychoanalytic Psychology, 32*, 94–115.

Maslow, A.H. (1987). *Motivation and personality.* (3rd Ed). London, England: Pearson Education.

Maslow, A.H. (1999). *Toward a psychology of being.* (3rd Ed). Malden, MA: J. Wiley & Sons.

McWilliams, N. (1974). Contemporary feminism, consciousness-raising, and changing views of the political. In J. Jaquette (Ed.), *Women in politics.* New York, NY: Wiley.

McWilliams, N. (1977). Personality theory at middle age: Generativity vs. stagnation. *Contemporary Psychology, 22*, 25–28.

McWilliams, N. (1979). Treatment of the young borderline patient: Fostering individuation against the odds. *Psychoanalytic Review, 66*, 339–357.

McWilliams, N. (1980). Pregnancy in the analyst. *American Journal of Psychoanalysis, 40*, 367–369.

McWilliams, N. (1984). The psychology of the altruist. *Psychoanalytic Psychology, 1*, 193–213.
McWilliams, N. (1986). Patients for life: The case for devotion. *The Psychotherapy Patient, 3*, 55–69.
McWilliams, N. (1987). The grandiose self and the interminable analysis. *Current Issues in Psychoanalytic Practice, 4*, 93–107.
McWilliams, N. (1991). Mothering and fathering processes in the psychoanalytic art. *Psychoanalytic Review, 78*, 525–545.
McWilliams, N. (1992). The worst of both worlds: Dilemmas of contemporary young women. In B. Wainrib (Ed.), *Gender issues across the life cycle* (pp. 27–36). New York, NY: Springer.
McWilliams, N. (1993). Sexual intimacy and the therapist's educative role. *New Jersey Psychologist, 43*(1), 17–19.
McWilliams, N. (1994a). Ongoing therapy with a dissociative patient. *New Jersey Psychologist, 44*(2), 13–15.
McWilliams, N. (1994b). *Psychoanalytic diagnosis: Understanding personality structure in the clinical process.* New York, NY: Guilford Press.
McWilliams, N. (1994c, December). Working sensitively with students in the borderline range. *Current Issues in Counseling and Psychotherapy*, 1–7.
McWilliams, N. (1996). Some reflections of a heterosexual female therapist on working with lesbian, gay, and bisexual clients. *Gender and Psychoanalysis, 1*, 203–221.
McWilliams, N. (1998). Relationship, subjectivity, and inference in diagnosis. In J. W. Barron (Ed.), *Making diagnosis meaningful: Enhancing evaluation and treatment of psychological disorders* (pp. 197–226). Washington, DC: American Psychological Association Press.
McWilliams, N. (1999). *Psychoanalytic case formulation.* New York, NY: Guilford Press.
McWilliams, N. (2000). On teaching psychoanalysis in antianalytic times: A polemic. *American Journal of Psychoanalysis, 60*, 371–390.
McWilliams, N. (2002). Foreword to P. Casement, *Learning from Our Mistakes* (pp. ix–xii). New York, NY: Guilford Press.
McWilliams, N. (2003). The educative aspects of psychoanalysis. *Psychoanalytic Psychology, 20*, 245–260.
McWilliams, N. (2004a). *Psychoanalytic therapy: A practitioner's guide.* New York, NY: Guilford Press.
McWilliams, N. (2004b, Winter). Some observations about supervision/consultation groups. *New Jersey Psychologist*, 16–18.

McWilliams, N. (2005a). Preserving our humanity as therapists. *Psychotherapy: Theory, Research, Practice & Training*, *42*, 139–151.

McWilliams, N. (2005b). Response to Norcross. *Psychotherapy: Theory, Research, Practice & Training*, *42*, 156–159.

McWilliams, N. (2006a). Depressive and masochistic personality organization. *Issues in Psychoanalytic Psychology*, *27*, 95–117.

McWilliams, N. (Assoc. Ed.) (2006b). *Psychodynamic Diagnostic Manual*. Silver Spring, MD: Alliance of Psychoanalytic Organizations.

McWilliams, N. (2006c). Some thoughts about schizoid dynamics. *Psychoanalytic Review*, *93*, 1–24.

McWilliams, N. (2006d). The woman who hurt too much to talk. *Fort Da*, *12*, 9–25.

McWilliams, N. (2007a). Freud's contemporary relevance. In J. P. Merlino, M. Jacobs, J. Kaplan, & K. L. Moritz (Eds.), *Freud at 150: Twenty-first century essays on a man of genius* (pp. 143–146). New York, NY: Jason Aronson.

McWilliams, N. (2007b). Freud's voice. In J. P. Merlino, M. Jacobs, J. Kaplan, & K. L. Moritz (Eds.), *Freud at 150: Twenty-first century essays on a man of genius* (pp. 169–171). New York, NY: Jason Aronson.

McWilliams, N. (2009a). Psychoanalysis. In I. Marini & M. A. Stebnicki (Eds.), *Professional Counselor Desk Reference* (pp. 289–300). New York, NY: Springer.

McWilliams, N. (2009b). Some thoughts on the survival of psychoanalytic practice. *Clinical Social Work*, *37*, 81–83.

McWilliams, N. (2010). Paranoia and political leadership. *Psychoanalytic Review*, *97*, 239–262.

McWilliams, N. (2011a). *Psychoanalytic diagnosis: Understanding personality in the clinical process* (rev. ed.). New York, NY: Guilford Press.

McWilliams, N. (2011b). *The Psychodynamic Diagnostic Manual*: An effort to compensate for the limitations of descriptive psychiatric diagnosis. *Journal of Personality Assessment*, *93*, 112–122.

McWilliams, N. (2012). Beyond traits: Personality as intersubjective themes. *Journal of Personality Assessment*, *94*, 563–570.

McWilliams, N. (2013a). The impact of my own psychotherapy on my work as a therapist. *Psychoanalytic Psychology*, *4*, 621–626.

McWilliams, N. (2013b). Psychoanalysis and research: Some reflections and opinions. *Psychoanalytic Review*, *100*, 919–945.

McWilliams, N. (2013c). Psychodynamic therapy. In L. S. Greenberg, N. McWilliams, & A. Wenzel (Eds.), *Exploring three approaches to psychotherapy: The next generation* (pp. xx–xx). Washington, DC: American Psychological Association.

McWilliams, N. (2014). Reflections on the effects of a Protestant girlhood. In G. S. Stavros & S. J. Sandage (Eds.), *The skillful soul of the psychotherapist: The link between spirituality and clinical excellence* (pp. 17–29). Lanham, MD: Rowman & Littlefield.

McWilliams, N. (2015). More simply human: On the universality of madness. *Psychosis: Psychological, Social and Integrative Approaches, 7*, 63–71.

McWilliams, N. (Assoc. Ed.) (in press). *Psychodynamic diagnostic manual* (2nd ed.). New York, NY: Guilford Press.

McWilliams, N., & Lependorf, S. (1990). Narcissistic pathology of everyday life: The denial of remorse and gratitude. *Contemporary Psychoanalysis, 26*, 430–451.

McWilliams, N., & Stein, J. (1987). Women's groups run by women: The management of devaluing transferences. *International Journal of Group Psychotherapy, 37*, 139–153.

McWilliams, N., & Weinberger, J. (2003). Psychodynamic psychotherapy. In G. Stricker & T. A. Widiger, *Comprehensive Handbook of Psychology, Vol. 8: Clinical Psychology* (pp. 253–277). New York, NY: Wiley.

Messer, S., & McWilliams, N. (2006). Insight in psychodynamic therapy. In L. Castonguay & C. Hill (Eds.), *Insight in psychotherapy* (pp. 9–29). New York, NY: Springer.

Messer, S. B., & McWilliams, N. (2003). The impact of Sigmund Freud and *The Interpretation of Dreams*. In R. J. Sternberg (Ed.), *The anatomy of impact: What makes the great works of psychology great?* (pp. 71–88). Washington, DC: American Psychological Association Press.

PDM Task Force. (2006). *Psychodynamic Diagnostic Manual (PDM)*. Silver Spring, MD: Alliance of Psychoanalytic Organizations.

Rogers, C. (Producer), & Shostrom, E. (Director). (1965). *Three approaches to psychotherapy (DVD)*. Corona Del Mar, CA: Psychological & Educational Films.

Rosenthal, H. (2006). *Therapy's best: Practical advice and gems of wisdom from twenty accomplished counselors and therapists* (pp. 221–238). New York, NY: Haworth Press.

Sacks, O. (1990). *Awakenings*. New York: Harper Perennial.

Schafer, R. (1979). On being a psychoanalyst of one persuasion or another. *Contemporary Psychoanalysis*, 15, 345–360.

Sohtorik, Y., & McWilliams, N. (2011). Hugging, drinking tea, and listening: Mental health needs of Turkish immigrants. *Journal of Multicultural Counseling and Development, 39*, 66–77.

Toronto, E. L. K., Ainslie, G., Donovan, M., Kelly, M., Kieffer, C. C., & McWilliams, N. (Eds.) (2005). *Psychoanalytic reflections on a gender-free case: Into the void.* New York, NY: Routledge.

World Health Organization, Geneva. (2004). International statistical classification of diseases and related health problems 10th revision, 2nd ed. Retrieved from http://www.who.int/classifications/icd/en/HistoryOfICD.pdf

Marie T. Hoffman, who holds a PhD, is Assistant Clinical Professor of Psychology (Adjunct) at NYU Postdoctoral Program in Psychotherapy and Psychoanalysis, Co-Director of Society for Exploration of Psychoanalytic Therapies and Theology, Director of Brookhaven Institute, and Co-Director/Clinical Psychologist/Psychoanalyst at Brookhaven Center in Allentown, Pennsylvania. In addition to journal articles and book chapters, she is author of *Toward Mutual Recognition: Relational Psychoanalysis and the Christian Narrative* (2011) and *When the Roll Is Called: Trauma and the Soul of American Evangelicalism* (2016).

FSC
www.fsc.org
MIX
Papier | Fördert
gute Waldnutzung
FSC® C083411

Kolibri 360 GmbH
Ferdinand-Jühlke-Straße 7
99095 Erfurt, Deutschland
produktsicherheit@kolibri360.de